MENTAL ILLNESSES AND TREATMENTS

BRAIN-BASED TREATMENT

A NEW APPROACH OR A WELL-FORGOTTEN OLD ONE? NEUROPSYCHOLOGY AND PSYCHOTHERAPY

MENTAL ILLNESSES AND TREATMENTS

Additional books and e-books in this series can be found on Nova's website under the Series tab.

MENTAL ILLNESSES AND TREATMENTS

BRAIN-BASED TREATMENT

A NEW APPROACH OR A WELL-FORGOTTEN OLD ONE? NEUROPSYCHOLOGY AND PSYCHOTHERAPY

VANYA MATANOVA
MARTIN KOLEV
PIA TOHME
AND
ZLATOMIRA KOSTOVA

nova science publishers
New York

Library of Congress Cataloging-in-Publication Data

ISBN: 978-1-53616-951-5
Library of Congress Control Number:2019956503

Published by Nova Science Publishers, Inc. † New York

CONTENTS

PREFACE

What is the influence of psychotherapy over the brain functioning? Is it possible for us to determine in which type of psychotherapy the most significant changes in the brain functioning are observed? If the influence of psychotherapy over the brain is a fact, does this fundamentally change the training in psychology, psychiatry and psychotherapy and the inclusion of knowledge from the basic sciences?

Psychotherapy can be considered as a type of training that stimulates the patient to learn how to change their behaviour, thinking and regulation of emotions. Psychotherapy is much more than an opportunity to talk to someone with good listening skills. Understanding the relationship between psychotherapy and brain functions is stimulating news and changes the traditional thinking about the place of psychotherapy and pharmacotherapy, the relations between them in the process of overcoming mental problems and diseases. Psychotherapy is assumed to be a form of learning which suggests that the uptake of information in the process of psychotherapy leads to a change in the expression of genes, thus changing the strength of synaptic connections. The gene sequence does not change under the influence of the environment, but the ability of genes to direct the synthesis of individual proteins depends on environmental factors and is regulated by their influence (Kandel, 1998). This explains the phenotypic differences between monozygotic twins and discordance of diseases, such as schizophrenia.

Any mental condition is a brain state and any mental disorder is a disorder in the work of the brain. The effect of treatment of mental disorders is associated with an effect on structural and functional changes in the brain (Kandel, 1998).

Neuroimaging gives an objective diagnostics of mental disorders, revealing the reasons for their occurrence and therefore allows for the development of more effective methods of treatment and psychotherapy.

Chapter 1

PSYCHOTHERAPY - PAST AND PRESENT

Vanya Matanova and Zlatomira Kostova

"Biological approach - that's what psychotherapy needs."

-Eric Kandel

It is well known that in most of the cases, our deepest experiences are not related to current events. They are a result of the activation of past events that have lingered in us for a long time. Our negative experiences could hardly disappear forever. Sometimes even long-term psychotherapy cannot attenuate the negative experiences and they keep on influencing our present lives. Can you imagine a person whose childhood lacks unjust punishment or a word from their parents or the teasing, mockery, and put-downs of the peers? Suffering is inherent to human beings. Lack of suffering does not make us happy. Whenever we suffer, we strive to find the reasons in order to reduce or neutralize the negative thoughts, emotions, or bodily sensations. When we try to escape from suffering, the pain from the present intensifies.

Psychotherapy has a long history. Its path is marked by epochal discoveries in the field of medicine. It has always been a shared field of several scientific fields. It was Z. Freud who called psychoanalysis 'a treatment through conversation,' which implies that this treatment has no direct impact on the psyche or the symptoms of mental disorders and illnesses. This means that the changes in a patient are not caused by the direct impact on their psyche, but are a result of a particular communication and situation of protection, which, in fact, is an impact on the surroundings and implies cultural dependence for each psychotherapeutic treatment. In 2000, Eisner talked about the 'death of psychotherapy' referring to numerous studies that claimed that contemporary psychotherapy had become an intellectual entertainment. According to him, it could, at best, claim to be an applied science, but not a fundamental science. As far back as 1994, Eysenck, H. J. expressed his belief that the only real theory in psychotherapy was the concept of I. P. Pavlov on conditional reflexes and his model of experimental neurosis, and all the other concepts of psychotherapy were speculative and not confirmed by scientific facts. By using the latest achievements in neuroscience, it is possible for psychotherapy to become a science that helps regulate human behaviour and mental abilities. Conte (2009) believes that it is possible to minimize the fundamental mistakes of psychotherapy - mistakes of omnipotence, mistakes of interpretation, mistakes in judgments, language and ethical mistakes.

In its essence, psychotherapy is all three: science, art and craft. Science, because of the scientific experiments to test reality. Reality, however, is not only what happens in the outside world but also what is in the mind and brain of both participants in the process. Fundamental knowledge in the fields of physiology, psychophysiology, biology of the mind, etc. is necessary to step on it. Art, because it is related to creativity, emotion, imagination, and intuition - all of them are used to create something new. This implies constant search and improvement of new approaches and methods tailored to the needs of the modern person and their place in our technological age. Craft, because it implies life experience, life-long learning, the ability to understand and create a therapeutic situation, contact, context, an atmosphere of understanding and safety, providing conditions for change in the patient. The psychotherapist must look and see, listen and hear, understand and support, hear behind the words, see behind the action, and look for the 'black box,' which in fact is orange! And all this in a situation where only the patient knows something about their own lives and only they can decide what to share and what not. This suggests that the psychotherapist is not an expert in the patient's life and hence their inner world is beyond psychotherapist's reach!

Amini (1996) suggests that psychotherapy represents new attachment relationships which help to restructure that part of the implicit, procedural memory that affects attachment. The prototypes stored in memory are changed under the influence of the new relationships with the psychotherapist, which are internalized by the patient. This model implies the emotional involvement of the psychotherapist due to the fact that implicit learning depends on the emotional experience of the psychotherapist.

Nowadays, psychotherapy is defined as 'the informed and intentional application of clinical methods and interpersonal stances derived from established psychological principles for the purpose of assisting people to modify their behaviours, cognitions, emotions, and/or other personal characteristics in directions that the participants deem desirable' (APA, 2013).

This suggests the existence of the therapist-patient dyad, and in this dyad, the patient talks about their problems and symptoms in a supportive environment. In fact, much more happens than that. Often, therapy focuses on recognizing and changing thought patterns and beliefs that cause suffering and problems with functioning and relationships. Sometimes, our thought patterns are so automatic and unconscious that we don't even notice them. How often do we think about our thinking? Thoughts often lead to feelings which, in turn, lead to behaviours. Or about our feelings? These specifics form the basis of the sense of omnipotence in the psychotherapist. In order to minimize this omnipotence, it is necessary to turn the dyad into a triad. The missing link in this triad is the objective data of the patient's brain functioning, which will objectify the effectiveness of psychotherapeutic impacts.

The results of psychotherapy are usually determined by assessing the changes at mental and social level by using the self-report of the patient and their close relatives, or as a professional assessment of the psychotherapist themselves. There is no doubt that every mental activity is carried out through brain mechanisms. This means that any change in mental processes is associated with changes in brain functions or structures. There is already evidence that our subjective experiences affect our brain. Psychotherapy is based on patient's subjective experiences and therapist's competence to assess patient's profile and resources. Objective data on patient's functioning are often missing or they do not show any deviations, but this fact in most cases does not change patient's experiences, and, on the contrary, it even complicates them due to patient's emerging belief that 'their disease' cannot be identified because of the therapist's inability or due to its exclusivity and complexity. Still, the understanding of mental problems is associated with a change in the functioning of individual organs - that is, the clinical model in understanding mental problems is still alive and sufficiently stable. The belief that negative experiences have some 'brain' reason turns out to be true, especially after the boom in development of neuroimaging technologies.

To assess the effectiveness of any treatment it is necessary to make a quantitative assessment based on reliable data. This helps to make informed therapeutic decisions and to outline the prognosis of the disease. Even ancient Greeks knew that diseases were the result of physiological disorders and, therefore, they were recognizable and subject to study, description, and treatment. Any disease can be characterized according to the observed symptoms, it can be studied and, depending on the patient's particularities, a therapeutic plan can be evolved and its outcome can be foreseen. According to the definition of Evidence-Based Medicine published in British Medical Journal in 1996, 'The practice of evidence-based medicine means integrating individual clinical expertise with the best available external clinical evidence from systematic research and the patient's unique values and circumstances,' i.e., the informed choice of the patient. In other words, when making a decision, evidence-based medicine makes place both for objective science and the patient's value system. Many difficulties accompany this transition from the paternalistic decision-making approach because it requires up-to-date scientific information, moral qualities, limitation of omnipotence and expert opinion.

Is it possible to talk about evidence-based psychotherapy? A number of meta-analyses have been carried out on the effectiveness of psychotherapy under controlled conditions, and patients with similar problems have been selected in the study groups. A great deal of data is being processed, and based on this, new knowledge is continually generated. Grawe K., (2007) derives many important ideas from this database:

- The therapeutic relationship, a.k.a. therapeutic alliance, is a key factor to effective psychotherapy,
- Positive discussion of the future is much more important than discussing the traumatic past,
- Taking into account neurophysiological reactions and regularities leads to much more effective psychotherapy.

Evidence-based psychotherapy is, indeed, brain-based psychotherapy.

Great Discoveries in Neurology and Their Implications in Psychotherapy

The basic postulate of psychotherapy is that the mind and brain are separate but interacting units. Psychotherapy claims that it is directed towards the mind (or psyche), but acknowledges that the effect of psychotherapeutic interventions is at least partly due to modifications of neuronal networks and, in particular, brain neuroplasticity.

The history of psychiatry and psychotherapy is related to its consolidation with neurology. The history of psychology, psychotherapy and neuroscience extends from the Ancient world to the Nobel Prize of Eric Kandel in 2000, John Gurdon and Shinya Amanaka in 2012, and John O'Keefe, May-Britt Moser and Edvard Moser in 2014. Their discoveries about brain mechanisms of memory and plasticity of the nervous system pose new challenges to psychiatry, clinical psychology and psychotherapy.

Many of the therapeutic tools in psychiatry have been adopted by psychotherapy, especially because psychotherapy is a shared territory with different medical and non-medical professions. The loss of identity and confidence in psychiatry has led to changes in modern clinical psychology and psychotherapy consisting of blurring the boundaries between them. Many methods are shared by both psychiatry and clinical psychology, as well as by psychotherapy. Disagreements and contradictions both in clinical psychology itself and psychotherapy, and in their relationship to psychiatry, are an inevitable consequence of the uncertainty of assessment and intervention procedures conducted in psychiatry.

Historically, there are two traditions that can be outlined in the long history of psychiatry and psychotherapy with regard to the mind-brain issue. The relationship between the brain and behavior is defined as strictly causal. There are a number of different assumptions about how the brain influences the mind and how the mind influences behaviour. In Cartesian dualism, the mind is perceived as a separate ontological type and it is assumed that it has representative properties responsible for behaviour.

A traditional belief associates the brain with the psyche and hence with behavior. Perhaps the most famous early theorist in this direction was René Descartes (1596-1650), who attempted to describe the corporeal basis of human passions, theorizing how uncontrolled passions could lead to unusual or excessive behavior.

Thomas Willis (1621-1675) is considered one of the founders of the anatomy of the nervous system. Together with his colleagues he discovered the so-called cerebral arterial circle in the human body (*Circulus arteriosus cerebri*) that supplies blood to the brain, and introduced the concept of reflex. He created the enumeration of the cranial nerves, which is still used nowadays. He was the first to describe *Corpus striatum*, the thalamus, the bridge of Varolius (pons Varolii) and *Corpus mamillare*. He was the first to notice the link between depression and diabetes, which was rediscovered three centuries later. Willis studied the brain and compared the normal brain with the anomalies he had found in patients with mental retardation.

Mesmeris (Friedrich Anton Mesmer) (1734-1815) created a new scientific theory and universal treatment by provoking crises in patients; he used these crises for diagnostics and a method of treatment. He discovered the report or the therapeutic relationship. According to him, psychotherapy was based mainly on hypnosis and suggestion. Mesmer believed he had discovered the existence of universal, physical fluids whose equilibrium or disturbance explained health or disease. Soon, this theory was forgotten, and researchers turned to psychological concepts, such as the power of the will (Marquis de Puységur, 1751-1825) or to the idea of mental powers and nervous energy. Until the end of the 19th century, the general opinion was that a deficiency of nervous energy caused diseases.

Johann Christian Reil (1759-1813) was engaged in the study of brain anatomy and is considered one of the founders of rational psychotherapy. He developed a triple system of management and treatment: administrator, physician and psychologist. He believed that psychotherapy should be based on a precise system of 'practical empirical psychology'. The method of treatment should be adapted to the patient's individual needs. Reil identified three groups of psychotherapeutic methods:

- Bodily stimulation aimed at changing the general physical condition.
- Sensory stimulation, for example, the method of "therapeutic theater".
- The method of 'signs and symbols' - training based on reading and writing, occupational therapy and art therapy.

In the last years of his life, Jean-Martin Charcot (1825-1893) believed that between the area of clear consciousness and the area of organic brain physiology there was yet another vast realm. The increasing knowledge of the laws of 'faith treatment' led to great therapeutic success. He will go down in history as the scientist who discovered the mental realm about which no one had ever suspected, and as one of S. Freud's great predecessors.

Brown-Sequard (1817-1894) formulated the so-called theory of 'dynamic processes' in the nervous system, according to which the stimulation of a section of the nervous system could lead to an impact on any of the other ones in the form of either 'dynamogenesis' (stimulation of functions) or 'inhibition' (reduction of functions). Psychiatrists began using these concepts for mental disorders, complementing them with others in the field of brain physiology (for example, the term 'facilitation' or 'relief').

Hughlings Jackson (1835-1911) talked about the evolution of the nervous system. According to him, some of the centers appeared at a later stage of evolution than others. The younger these centers are, the more vulnerable they are. When one of them is damaged, the activity of the ancient ones increases.

The term 'psychotherapy' became fashionable around 1890 when the need for new treatment to meet the intellectual needs of high society patients had already arisen.

The Age of Enlightenment had a strong influence on psychiatry; it was the time when the process of its secularization began. Symptoms previously explained by magic began to be perceived as manifestations of mental illness. The rapid development of mechanics and physics led to the adoption of mechanistic models in physiology; mental life was considered an effect from nervous system activity, that is, as disorders of the mind. The causes of these disorders were perceived as physical damages, especially brain damages or a consequence of uncontrolled emotions. During the Enlightenment period, many attempts were made to link brain research to mental health problems and diseases. But the lack of empirical data hampered the explanations of mind-brain relations and they continued to be described and explained by philosophy.

In the first decade of the 19th century, there were two directions in psychiatry - somatic and mental. The somatic direction attributed the manifestations of mental disorders to physical causes and the state of the brain, while the mental direction attributed them to emotional states. At the end of the 19th century, there were attempts to unite the two trends, which led to the formation of two approaches in the treatment of mentally-ill patients. Pinel's humanitarian approach (Philippe Pinel, 1745-1826) required moral treatment of patients. The other approach was based on the understanding that 'mental disorders were diseases of the brain,' so it was necessary to study the anatomy and pathology of the brain in order to find a way to treat these conditions. Wilhelm Griesinger (1817-1868) declared war on the old romantic style in psychiatry.

Meanwhile, Carl von Rokitansky (1804-1878) and Rudolf Ludwig Karl Virchow (1821-1902) laid the foundation for cellular pathology. Soon

Theodor Hermann Meynert (1833-1892), Carl Wernicke (1848-1905) and their followers attempted to place psychiatry on that base. T. Meynert and C. Wernicke studied the anatomy of the brain and set up a common system of organic and mechanistic psychiatry. They researched the effects of brain diseases on speech and language. Wernicke noticed that not all language deficits were caused by damage in the Broca's area. He came to the conclusion that injuries in the left superior posterior temporal gyrus caused linguistic deficits known today as Wernicke's Aphasia. At the end of the 19th century, a number of psychiatrists used terms from brain anatomy to define psychopathological disorders.

T. T. Meynert suggested that the brain areas, which were phylogenetically more ancient, were the center of involuntary movements and were controlled by the cerebral cortex that developed later in evolution. He distinguished a 'primary ego' - formed as a result of the immediate functioning of the centers of the cerebral cortex, and a 'secondary ego', which was the product of the activity of a set of associative links. T. Meynert believed that when the activity of newer centers was impaired, the activity of phylogenetically older centers became increasingly important. Thus, he explained the origin of delusions of persecution and grandeur and treated these delusions as a mental manifestation of two basic instincts - attack and defense.

Pierre Briquet (1796-1881) included in hysteria sociological 'or material' phenomena, such as living and working conditions, natural circles and even the movement of the stars. He defined hysteria as 'neurosis of the brain, the manifestation of it consisting chiefly in a perturbation of those vital acts which are concerned with the expression of emotion and passion'. Briquet regarded it as 'dynamic suffering of those parts of the brain that determine affects and sensations.'

Emil Kraepelin (1856-1926) can be credited with overcoming these trends - he introduced a new approach to psychiatry that distinguished neuroscience, brain anatomy and experimental psychology. Kraepelin regarded the causes of mental disorders as predominantly biological by nature. Physiological factors and dysfunctions of body organs became the significant factors and not psychological reasons cherished by many representatives of the French school at that time. Auguste-Henri Forel, 1848-1931, holds a significant place in the history of neurology; he is famous for his discoveries in the field of brain anatomy, for being the co-founder of the Neuron doctrine and one of the first scientists to contribute to sexology.

Paul Eugen Bleuler (1857-1939) believed that schizophrenia occurred for an unknown reason, for example, the effect of a toxic substance in the brain. He distinguished between physiogenic symptoms resulting from an organic process and psychogenic symptoms - which were a consequence of the first.

Korbinian Brodmann (1868-1918) is the founder of the cytoarchitectonics of the cerebral cortex. He studied the cellular characteristics of cerebral hemispheres and split the cortex into 52 distinct regions based on their cellular characteristics, the so-called Brodmann's cytoarchitectural areas. Brodmann's followers discovered many other areas of the brain. Nowadays, thanks to functional magnetic resonance tomography, the existence of 180 cytoarchitectonic fields has been confirmed, 97 of them were discovered for the first time. New data helps explain the evolution of the cerebral cortex as well as the role of different areas. Prior to these discoveries, Brodmann's map was most widely used and until the early 1990s, it was not subject to serious revisions. Neuroimaging allows us to see brain activity in particular regions during a performance of various tasks, which raises the question of whether morphological differences determine the functions of the individual areas. According to Brodmann's map, speech areas are 22 and 44 - Broca's and Wernicke's areas. New technologies, however, show that speech and language areas are 6, 13, 20, 37, 38, 39, 40, 44, 45, 46, 47, which means that morphologically different fields can work together to perform a single function. For a long time, Brodmann's field 46 was mainly associated with working memory, but new results suggest that this field is also involved in securing the recombination of particular motor acts and formation of new action schemes, as close as possible to what is demonstrated by the other.

This confirms the idea that compensatory abilities and brain plasticity are not determined by the morphology of particular cortical areas, which necessitates the rethinking of therapeutic methods in cortical lesions.

Alexander B. Campbell (1788-1866) suggested cortical fields to be divided into primary, secondary, and tertiary ones. Primary and secondary fields receive impulses directly from the thalamus, while the tertiary ones from the primary and secondary fields. Primary fields perform a specific analysis of the impulses from a certain modality. Secondary fields interact between different analytical areas, while tertiary ones play a crucial role in complex types of mental activity - symbolic, speech, and intellectual activity.

According to Pierre Marie Félix Janet (1859-1947), hysteria and neuroses were caused by depression and brain exhaustion, which in turn is related to the narrowing of consciousness and the enhancement of suggestibility. He emphasized that 'the easier the treatment seems to be, the more severely the patient's brain is affected,' since a high degree of suggestibility indicates weakness of thinking. 'Maybe, someday we would be able to create a balance sheet of the energy reserve. Psychiatrists will then be able to effectively use the patient's poor resources by avoiding unnecessary costs and directing efforts where there's a need to do so and, even better, if we could teach the patient how to increase the amount of energy available to make their psyche more powerful'.

P. Janet never doubted the physiological nature of mental powers and he also believed that there would be a time when these forces could be measured with precision. He believed that these forces were to a large extent related to the state of the brain and the other organs and were subject to different trends. Every trend had a certain charge of mental energy, and mental powers could, in a certain way, be recovered and stored. 'I do not know where these reserves are, but I know they exist.' In his opinion, 'complete and full acts increase mental tension, while the unfinished acts that have not yet reached the goal decrease it.' Here is what he recommended:

- provide a level at which the patient can perform complete actions.
- make the patient perform and bring to an end tasks of a similar nature until the moment these tasks do not pose any difficulty to the patient.

- switch between tasks of varying difficulty, more complicated tasks and of a higher level.
- find other varied psychological investments.

Modern discoveries in neurophysiology confirm the presence of P. Janet's concept of the mental. Janet. P. Janet's concepts of vigilance and presentification have a substrate in some parts of the diencephalon.

Lashley (Brain Mechanism and Intelligence, Lashley, 1929) found that memories were stored in all areas of the cortex and their intensity depended on the total amount of active cells. He introduced the idea that the excitation of millions of neurons formed stable interfering patterns scattered all over the cortex and representing the basis for all information in the systems of perception and memory.

In the joint work of Sigmund Freud (1856–1939) and Josef Breuer (1842-1925), 'Studies on Hysteria', 1895, therapy was perceived as an adaptation of treatment through catharsis, which was analogous to P. Janet's treatment. It is impossible to understand the system of psychoanalysis without analyzing the scientific trends in the last decades of the 19th century. Freud's psychotherapeutic method has been developed over an extended period; it has undergone several metamorphoses, led by Freud himself and by his students.

One of the tendencies which influenced S. Freud was the denial of metaphysics and the tendency to study the world from a scientific point of view. At that time, however, science had a lot of limitations, including technological ones, which formed the understanding that part of reality was unknowable. According to Henri Frédéric Ellenberger (2001), denial of metaphysics led to the fact that the soul was left outside the subject of psychology, and vitality - outside the field of biology. Neurophysiologists tried to explain mental processes with brain structure, through physical and chemical processes.

During his practice as a neuroscientist, S. Freud made histological studies of the medulla oblongata in T. Meynert's laboratory (Meynert 1883-1886), who was considered one of the greatest brain anatomists. On January

21st, 1885, S. Freud applied for the post of Privatdozent /associate professor/ in neurology. His clinical practice involved ophthalmology, dermatology and pediatric training.

While working as Head of the Neurology department at Kassowitz's Institute (M. Kassowitz, 1842-1913), Freud collected very rich clinical data. One of his first studies was on the structure of the gonads in eels. Freud spent a lot of time exploring the cells from the spinal cord of some fish species, and after that, he studied *corpus restiforme* and the auditory nerve cells. These were valuable studies some of which he shared in articles ('The structure of the elements of the nervous system'). Today these studies are regarded as predecessors to the Neuron doctrine. He used the anatomo-clinical method to check the clinical diagnoses from the autopsy. His achievements in the classification of cerebral palsy types, as well as aphasia and agnosia types, were also remarkable. For these 20 years, Freud went a long way from microscopic anatomy through anatomo-clinical neurology to his 'Project of Scientific Psychology'.

'Project of Scientific Psychology' was part of the correspondence between Freud and Wilhelm Fliess in the period from 1895 to 1930 and was published in a separate edition in 1950. The 'Project' is generally considered a version of reductive theories of mind and belongs to the classical heritage of psychoanalysis. Long after Freud's death, this project was underestimated by the different psychoanalytic traditions.

To a certain degree, the basic concept in the 'Project' is contradictory, as it draws closer to the outdated concept of 'epiphenomenalism', in which a person's mental life is seen as an accompanying by-product of brain activity. The main idea in the 'Project' is the correlation between mental processes and distribution and circulation of the energy through certain layers of matter, i.e., hypothetical brain structures.

According to Freud, the quantity of energy is equal to the sum of excitations arising either in the outside world through the sensory organs, or in the internal one, i.e., in the body. The quantity of energy is driven by two principles - the principle of complete energy discharge and the principle of sustainability that keeps the sum of excitations a constant. According to Freud, material elements are neurons, divided into three types:

- "phi", they receive excitation from the outside world and are driven by inertia,
- "psi", they receive excitation from the body or neurons type «phi» and are driven by the principle of sustainability, they keep the traces of each registered irritation and form the basis of memory.
- "Omega", they receive excitation from the body and from neurons of the «phi» type; they have the ability to transform quantity into quality due to the period of movement. They form the basis of perception.

According to S. Freud, dissatisfaction is connected with an increase in the level of excitation, while pleasure is connected with a discharge of excitation. Ego is, in fact, a neuronal network organization responsible for the permanent stock of excitation and is capable of forbidding the uptake of excitation. These ideas of S. Freud can be traced back to the works of Johann Friedrich Herbart (1776-1841) and Gustav Theodor Fechner (1801-1887).

The 'Project' from 1895 can be regarded as a logical continuation of the theories of Freud's predecessors. Many of their ideas would appear in a different form in his psychoanalytic theory. In fact, S. Freud was the first who tried to establish the relationship between brain and mind in his 'Project of Scientific Psychology'. He talked about the synapse, although he called it a 'contact barrier', he described how a synapse could be changed under the effect of learning, thus predicting and provoking the work of E. Kandel. Every founder is always the successor of previous ideas and forerunner of further discoveries.

There is yet another forgotten statement by S. Freud that poses many questions to contemporary psychotherapy. He argued that we did not know what lay between the psyche and the brain, but he predicted that there would be times when all our psychological assumptions would get an organic basis.

S. Freud, as a man of great scientific and literary awareness and culture, lived and worked on the border of several cultural trends, he was able to quickly perceive and process new ideas, to integrate them into his observations and views. Quite a few of his ideas were known and shared by his colleagues, and to a certain extent, they were the forerunners of S. Freud's psychoanalysis. His extraordinary knowledge in various fields of medical science, as well as his advanced knowledge in fundamental research, provided him with a basis solid enough to form a new psychotherapeutic method, a system of his own. He passes the path from microscopic anatomy to theoretical neurology and hence to psychiatry and psychoanalysis. Maybe this is the beginning of translational research? The knowledge of this first period of his professional development is important for the understanding of psychoanalysis.

He believed that one day we would be able to explain any psychopathology with brain functions, but until then we would need independent mental constructs to explain the etiology of such pathologies. Freud did not describe the psyche as a separate ontological unit but argued that due to its representative nature, the mind could be discussed in a way independent of body physiology. It turns out that he knew about the relationship between psychopathology and brain functioning, but since at that time there was no proof of it, he created a system to work with until this relationship was proven.

Contemporary Theories and Their Place in Psychotherapy

The modern development of psychotherapy is influenced by some theories that arose in the XX and the early XXI centuries. Small portions of these theories have been integrated into psychotherapeutic methods for the treatment of various mental disorders, but they have not found their systematization and application in modern psychotherapy.

The first studies related to 'theory of mind' appeared 30 years ago. The difficulty in these studies is connected with the fact that 'theory of mind' lies at the crossroads of many sections of psychology, philosophy, and cognitive studies. This area is quite varied, but it is the ontogenetic study of 'theory of mind' that generates most interest. This requires an answer to the question of how the ability to understand and adjust one's behavior to the mental state of others changes, what the reasons for the development of this ability are and how it affects general cognitive abilities. 'Theory of mind' is a system of representations of mental phenomena - meta-representations, which develop intensively in childhood. To have such a system means to be able to perceive not only your own experiences, knowledge, intentions, beliefs, etc., but also the beliefs of others, which allows you to explain and predict their behavior. An important aspect of this system is the perception of the other as an intentional agent or the acceptance and understanding of the fact that one's mental state is not always the same as the mental state of the other. (Goldman, A. I., 2012).

'Theory of mind' refers to our ability to ascribe independent notions to ourselves or others in order to explain our own or somebody else's behavior. Such notions need to be independent of both the real state of affairs and the notions of others. In other words, 'theory of mind' can only be manifested in the explanation and prediction of behavior that harbors false expectations, since if attraction of realities is required to explain behavior, reasoning about the notions of others is not at all required.

'Theory of mind' stems from the impossibility of directly observing the mental processes of others, i.e., we have no direct access to the mental world of others, which means that the assumption of the existence of mind in others is theoretical. Everyone assumes that others have thinking analogous to their own based on such indications as two-sided social interaction, shared attention, communicative skills and understanding of others' emotions and actions. 'Theory of mind' allows us to imagine the thoughts of others, their feelings and aspirations and to guess their intentions. This means that intra-mental processes are the reason for the behaviors of others and this is the way in which we can have any predications along these lines. The ability to ascribe different thoughts and feelings to others as a reason for their behavior allows us to consider the mind as a generator of representations. 'Theory of mind' is a natural ability, but an experience of social interaction is needed for its development. Empathy is close to this concept and is related to the ability to recognize and immediately experience the likely desires, beliefs and feelings of others. 'Theory of mind' is not limited to one level only, and, in this sense, the representations of internal notions can grow infinitely.

Initially, it was assumed that 'theory of mind' is formed not earlier than at the age of three or four, but it is still arguable whether or not 'theory of mind' exists in children under the age of three. It is difficult to establish because of the still undeveloped speech, which prevents the assessment of a child's internal representations. Empirical data show that children under the age of four cannot cope with false belief tasks but cope fairly well with their own false beliefs.

A number of empirical studies reveal that manifestations of such ability can be noticed even in younger children. A. Meltzoff (2003) studied the understanding of intentions of others in 18-month-old babies. They understand our aims, even when we are unable to fulfill them. The results of these studies busted the myth about babies' antisociality. An issue that is increasingly emerging is related to the problem of creating a common theory for the development of 'theory of mind', which could systematize and clarify the terminology and methodology, as well as the phenomenology and chronology of 'theory of mind'. It is necessary to clarify the connection between the symbolic and the social. Traditionally, it is assumed that speech is the link between them. But symbolic play, or make-believe play, is not related to speech. It is triggered by imagination and, in turn, it triggers imagination, thus provoking the ability to detach from the existing situation, the ability to imagine something to be something else, despite the sensory experience. This is probably the same ability that triggers abstract thought. That brings into question the understanding that society creates the symbolic through speech. Speech formation also requires social ability and the ability to symbolize. However, it may be assumed that speech development stimulates the ability to symbolize, but this ability is independent of speech, although speech is based on it. It is well known that an intentional environment stimulates speech development, which leads to the development of symbolic operations. It appears that young children are also more interested in intentions than in words. Maybe this 'blindness' of literal meaning should be seen not as immaturity, but as an advantage of communication over language.

'Theory of mind' and 'mind-reading' are seen as components of an affective reaction that is part of the 'Theory of Blind Consciousness'. In that way, not only the mental state of the other is determined, but also it makes it possible to predict their emotional state and to react to it with a corresponding emotion. Mind-reading and mentalization help us not only figure out the behavior of others but also allow for a whole set of mental states to help us perceive possible follow-ups in the behavioural patterns of others. This allows us to understand the meaning of the other's behavior, to predict what they will do at a certain moment and how they will feel.

'Theory of mind' relates to the ability of the person to understand the state of mind of others, which differs from his or her own. It is impossible not to notice the proximity of this term to constructs such as reflection, empathy, self-awareness, metacognition, etc. The choice of the term 'theory of mind' is not accidental. To a large extent, this choice is determined by the empirical complexity of the constructs that are difficult to verify. The mystery of this phenomenon lies in the fact that this state is acquired differently. It is difficult to answer the question of how we get to the knowledge that others experience states similar to our own, although the same situation may trigger different experiences among different people.

After the initial enthusiasm that accompanied 'theory of mind', a number of questions emerged, which did not find their answer in it. This gave a considerable impetus to conducting various empirical studies aimed at clarifying the mechanisms for the development and the perceived deficits of 'theory of mind'. 'Theory of mind' is related to the ability to understand the whole range of mental states - beliefs, desires, intentions, imagination, emotions, etc. which are the cause of action in ourselves and others. 'Theory of mind' is used to denote the unconscious ability to understand the inner states of others. Awareness is a very important aspect of this theory. Self-awareness is actually a consequence of the awareness of the psyche. There is no reason to believe that there is a difference in the awareness of the mental state of others and the awareness of one's own state.

Theory-theory is a concept for the organization not only of the 'theory of mind' but also of knowledge in general. Prediction, explanation, and interpretation are made following the internal structure of knowledge. The child is actually a scientist who develops, by analogy, with the evolution of science. There are three stages in the development of 'theory of mind'.

1. Conceptualization of desires, emotions, and cognitive experience. This conceptualization is not representative; desires are related to objects from the real world - the desire is present and it dominates, while the inner notion is absent (at about two years of age).
2. Awareness of the presence not only of desires but also of beliefs, where desires explain actions. Children gradually begin to realize that beliefs as internal representations may turn out to be false (at about three years of age).
3. Children imagine that desires and beliefs define behavior and that beliefs are an interpretation of reality (at about four years of age).

According to the Simulation theory of empathy, children are aware of their mental states and can form conclusions about the mental states of others by way of imitation. In the task of understanding false beliefs, the child can predict the beliefs of another person imagining that he or she is in their place. This process may be unconscious. Thus, 'theory of mind' is related to improving the ability to imitate. It is necessary for mastering cognitive knowledge and social skills, with experience playing a crucial role.

Modular theory has emerged in analogy with the modular theories of language and perception. In these theories, there is postulated a special mechanism in the brain which determines the mental states of people; it may be congenital or it may appear in the early stages of development. In other words, the development of 'theory of mind' in ontogenesis is determined by the neuronal development of a given module, *Theory of Mind Module* (ToMM), in which there are actually two submodules – one for agents and one for objects. The most active sections of the brain associated with 'theory of mind' are the prefrontal cortex and the border of the prefrontal premotor cortex.

Imitation is a field that is believed to play an important role in laying the foundations for subsequent social development. For example, imitation may be the basis of emotional reflection - employing sharing or assuming the same facial expression, the baby can learn to share the emotion experienced by the other. The lack of early imitation is more likely to have an adverse effect on further social development. Rogers and Pennington (1991) also suggested that the basic biologically determined deficit in imitation could be the main reason for the later absence of affective sharing with the main caregivers. On the other hand, Charman and Baron-Cohen (1994) found intact gestures and procedural imitation in autistic children. The question remains unclear whether neonatal imitation refers to the same cognitive abilities as later conscious imitation.

The Relevance theory of D. Sperber and D. Wilson (Sperber, D., & Wilson, D., 2004) attempts to explain communication between people. According to these scholars, cognitive processes are responsible for benefit/loss analysis – the achievement of maximum effect with minimum losses. This allows us to explain the fact that we are able to understand what others mean when they communicate with us or send us a non-verbal signal. In order to fully understand what has been said, and not just its literal meaning, the listener must figure out the speaker's intentions. It is closely related to 'the theory of mind' since conclusions about the speaker's and listener's mental states are very important components of successful linguistic interaction. Meta-representations are connected with the ability to distance ourselves from our perceptions and memory. This is possible through forming second-order concepts that refer to first-order ones. The difference is in the 'perception of something' and the 'knowledge about the perception of something'. Second-order concepts are of significant importance for make-believe play/pretending, imagination, social relationships, understanding the intentions, which are at the base of each communication. There are first-order representations that are connected to reality, while those of the second-order are representations of first-order representations. Second-order representations are further removed from reality and may exist together with first-order representations. Thus, there is no contradiction between the recognition that the cup is empty and the make-believe play when we pretend it is full. This 'pretense' is proof of a thought process that would be impossible without second-order representations.

Chapter 2

MENTALIZATION

Pia Tohme and Martin Kolev

THE DEVELOPMENT OF THE CONCEPT OF MENTALIZATION

Mentalization has been defined as a form of imaginative mental activity, perceiving and interpreting human behaviors as conjoined with intentional mental states such as needs, desires, feelings, beliefs and goals (Allen, 2008; Allen & Fonagy, 2006; Fonagy, Target, Steele, & Steele, 1998). It incorporates both, an affective and cognitive aspect (Slade, 2005), in other words thinking about feeling and thinking about thinking (Fonagy, Steele, & Steele, 1991; Fonagy & Target, 1997). This capacity develops in the child through interpersonal interactions with more mature minds, based on the quality of attachment relationships with caregivers. It is argued to reflect the extent to which the child's subjective experiences are appropriately mirrored (Fonagy, Bateman, & Luyten, 2012; Fonagy, Gergely, Jurist, & Target, 2003). Although the first explicit formulation of mentalization, as defined by Fonagy and his colleagues, was made in the nineties, the concept of mentalizing has its roots in classical psychoanalytic theory (Bouchard & Lecours, 2008; Fonagy, 1999).

THE ROOTS OF MENTALIZATION IN PSYCHOANALYTIC THEORIES

In its widest sense, mentalizing includes a process of transformation, echoing Freud's concept of Bindung/binding or linking (Freud, 1911). It is an ego function which transforms immediate physical quantities, or somatic experiences, into associative psychical ones, to restrict the free flow of excitations. This process occurs by linking ideas to one another, thus creating more stable forms, leading to the creation of associative pathways, part of secondary processes, in order to adapt to the external reality, by creating stable mental representations of the self and others (Freud, 1911; Laplanche & Pontalis, 1973; Lecours & Bouchard, 1997).

Bion's (1962) theory of containment also provides a basis for the concept of mentalization. He first pinpointed this activity in patients who were expressing ideas they could not fully understand themselves, thus needing a container, found in the therapist, in order to make sense of them. Bion (1962) then applied this idea to the mother-infant relationship, suggesting that the baby receives raw sensations from the outside which he/she cannot fully cope with on the inside, thus using the mother as a container. This was only deemed possible if the mother was in a state of *rêverie*, allowing her to tolerate the raw sense impressions, β-elements, projected onto her by the child, transforming them into α-elements, more tolerable to the baby, making sense of feelings of the self and others. Furthermore, the α-function plays a crucial role in the establishment of the contact barrier, differentiating between unconscious and conscious thinking, a notion echoed in the concept of mentalization, given that a pre- requisite to its acquisition is the ability to differentiate between reality and fantasy (Bion, 1962; Holmes, 2006). Only through this process can the child give meaning to raw emotional experiences. Through this continuous state of coming-to-know and repeated interactions with the caregiver, the child will, in time, internalize this function and regulate his own negative affective states (Bion, 1962; Fonagy, 1999; Fonagy et al., 2003; Holmes, 2006), or in other words, develop the capacity to mentalize.

The Winnicottian model also emphasized the transitional role played by the mother, necessary for the development of a sense of self in the baby (Winnicott, 1956). The mother is perceived as a holding object, similar to Bion's (1962) container, providing psychical cohesion to the infant. The mother, therefore, allows for the child's actions and gestures to be reflected to him/her through her expression (Bouchard & Lecours, 2008). The 'good enough' mother should consistently mirror the infant's needs by having the "capacity to put herself in the baby's place and to know what the baby needs" (Winnicott, 1962, p. 57), therefore keeping away the "unthinkable anxiety" experienced by the infant (Winnicott, 1962). It is argued that in normal development, the mother identifies with the child, allowing him/her to see him/herself in a way that validates the self. Thus, the infant begins to build up a distinct picture of both internal and external reality based on the perception of himself as thinking and feeling through his mother's mind (Fonagy et al., 2003; Holmes, 2006; Winnicott, 1956), a hallmark of mentalization.

Within the context of attachment theory, Bowlby (1969) revolutionized the way the tie between mother and infant was viewed. He suggested that it is through repeated dyadic interactions with the mother that the infant develops mental representations, or internal working models of the mind (IWM), which consist of a set of expectations of how the mother will respond to his/her needs. The infant, therefore, constructs IWM of the self as an independent being with feelings, goals and interests that are separate from the mother's (Bowlby, 1969; Bretherton, 1992). According to Bowlby (1973), these models of the self are derived from the way the child believes he/she is perceived by the attachment figure, the availability and responsiveness of the caregiver and the child's perception of the parents' accessibility in times of distress and need. Bowlby (1969) and Ainsworth et al. (1978) set these parental features as the central determinants of the child's attachment security, crucial to the development of mentalizing capacities.

Independently, a group of French psychoanalysts elaborated their own notion of mentalization, based on Freud's early concept of "binding" and the economic model of the mind (Lecours & Bouchard, 1997; Luquet, 1988; Marty, 1968). Marty (1968) considered mentalization as ensuring stability, as he suggested it plays the role of a protective buffer in the preconscious system in order to prevent disorganization within the systems. He suggested that mentalization allows fluidity in the use of associations linking drive excitations and internal representations (Marty, 1990; 1991). Luquet (1987) distinguished different forms of thinking and organization of inner experience, including primary mentalization and secondary mentalization. The former was characterized by the absence of mentalization while the second was conceptualized as symbolic mentalization, linked with sensory data and primary unconscious fantasy as seen through dreams or play. Summarizing the French definition of the concept of mentalization, Lecours and Bouchard (1997) explained that it can be conceived as the antithesis of *pensée opératoire*, i.e. thinking devoid of affect. They stated that mentalization is, firstly,

"A process of transformation, [...] a preconscious/ego activity that transforms, maintains and further elaborates basic somatic or motor drive-affect experiences into psychic contents. This transformation is achieved through a linking activity that establishes representations and symbols in order to permit the individual to free him/herself from the concrete and absolute nature of the primary motivational (drive-affect) pressures. Secondly, mentalization refers to a theoretical hierarchy of levels of psychic elaboration that differ qualitatively" (Lecours & Bouchard, p. 860).

In sum, psychoanalytic theories posited that a transformation process is necessary to move from raw and concrete experiences to recognizing and naming their equivalent affective states (Bion, 1962). This transformation was hypothesized to allow the differentiation between internal and external reality as well as between the mother's and the infant's feelings and goals (Winnicott, 1956; 1962). Within the context of attachment theory, Bowlby (1969) suggested that it is through repeated dyadic interactions with the mother that the infant constructs internal working models of the self as an independent being (Bowlby, 1969; Bretherton, 1992). It can therefore be argued that psychoanalytic thinking, focusing on mirroring, naming and containing of affect, underlies the affective component of mentalization, whereas attachment theory is more in line with studies looking at the cognitive aspect of mentalization described next.

THEORY OF MIND AND MENTALIZATION AS A COGNITIVE CAPACITY

Mentalization includes a cognitive capacity, rooted in the concept of theory of mind. Dennett (1987) proposed three stances available to predict behavior: 1) the physical stance, through which the individual understands behavior only through physical properties, 2) the design stance, which emphasizes the programming behind development and finally, 3) the intentional stance, which allows the individual to predict what the most rational move is in terms of a theory of mind.

Premack and Woodruff (1978) defined the concept of theory of mind as the ability to ascribe feelings and thoughts to the self and others, as well as to anticipate their influence on behaviors. This was related to their finding that children around 3 or 4 years of age start to understand the separateness of mind between self and other, and therefore recognize that thoughts, beliefs and feelings differ according to the knowledge each person possesses (Premack & Woodruff, 1978).

Researchers attempting to investigate the determinants of the development of theory of mind in children agree upon the importance of social and family factors, such as family structure (Hughes & Dunn, 1998; Jenkins & Astington, 1996), parental talk about emotions (Denham, Zoller, & Couchoud, 1994) and quality of parental control (Dunn, Brown, Somkowski, Telsa, & Youngblade, 1991). However, theory of mind alone was not found to adequately explain how children acquire the capacity to perceive mental concepts (Fonagy & Target, 1997); it was deemed insufficient in taking into account how the understanding of behaviors in terms of mental states is transmitted from the social world to the child (Fonagy & Target, 1997). Moreover, the theory of mind model suggests that mental states understanding comes from introspection, which has been criticized for focusing on conscious motivations rather than affect regulation and unconscious processes (Fonagy & Target, 1997). These limitations have led to a more complex conceptualization of mentalization as not only including a cognitive component, such as theory of mind, but also incorporating an affective component (Fonagy et al., 1991; Meins, 1999, van Ijzendoorn et al., 1995).

MENTALIZATION AS AN INTERACTION BETWEEN AFFECTIVE AND COGNITIVE COMPONENTS

The first explicit formulation of mentalization including both the affective and cognitive components was based on Marty's (1991) observation of clinical somatization disorders which he identified as devoid from any affective understanding, the antithesis of mentalization (Bouchard & Lecours, 2008; Holmes, 2006; Marty, 1991). Some philosophers of the mind have presented the idea of the theory of the *unconscious* mind (Hopkins, 1992; Wollheim, 1995) in an attempt to expand the concepts presented by Dennett (1987) and Premack and Woodruff (1978). They highlighted that behaviors should not solely be understood in terms of an interconnected set of beliefs and desires as explained by Premack and Woodruff (1978), but the unconscious mind should also be explored as reflected by unconscious mental states. This promotes the reach of a more comprehensive understanding of feelings and desires underlying behaviors (Fonagy & Target, 1997). Slade (2005), writing from a psychoanalytic perspective, summarized this capacity as including both a cognitive process, perspective taking (allowing the person to understand what the other person believes), and an affective process, through which one can regulate and contain one's own emotions and those of others.

More recently, developmental social and cognitive researchers Fonagy and Luyten (2009) considered mentalization as a multidimensional construct. They proposed that mentalization can be characterized as "organized along four polarities: 1) cognitive versus affective, 2) self- versus other-focused, 3) internally versus externally based and 4) automatic/implicit versus controlled/explicit. Each of these dimensions reflects the involvement of two relatively distinct neural systems" (Fonagy & Luyten, 2009, p. 1358).

Discussing the cognitive/affective polarity, they argued that, cognitive processes in mentalization rely on the same mechanism called upon during theory of mind tasks, based on attitude representations of the self and others. In contrast, affective processes in mentalization rely on the empathy system and representations of one's own and others' emotions, whereby representations of the other's emotions are based on the assumed effect of these emotions on the self (Fonagy & Luyten, 2009).

The second dimension distinguishes between one's self-reflective capacities, different but related to one's capacity to reflect about others. Lecours and Bouchard (1997) illustrated this using the example of a misunderstanding between two friends, whereby one should hold capacities on both extremes of this polarity (mentalizing the self and the other) in order to reach a solution. This taps into the idea that mentalizing reflects the capacity to acknowledge and make sense of the dynamic shift in feelings and intentions based on the specificity of an interpersonal situation.

The third polarity taps into the use of both internal features, thinking about thoughts and feelings, as well as external features, paying attention to external cues reflecting mental states of the self and others.

Lastly, explicit mentalizing is theorized as more conscious and controlled, a direct response to being asked about mental states, whereas implicit mentalizing occurs more automatically, such as intuitively mirroring others' affects, naturally holding the other in mind based on internal and external cues. Despite this process being automatic and out of our awareness, Allen (2003) emphasized that mental effort is implicated in being emotionally attuned and available.

THE NEUROBIOLOGY OF MENTALIZATION

As previously argued, mentalization is not an innate capacity; rather it is conceived as a developmental task, closely dependent on one's early attachment relationships and the caregiver's mirroring of the child's affect (Fonagy et al, 1991). Research has focused on looking at the biological basis of parental mentalization, emphasizing the importance of oxytocin, usually present in significant levels around childbirth and breastfeeding, times when the mother is increasingly psychologically attuned to the child (Buchheim, Heinrichs, George, Pokorny, Koops, Henningsen, O'Connor, & Gundel, 2009; Strathearn, Fonagy, Amico, & Montague, 2009). Oxytocin was found to facilitate mothers' capacity to markedly reflect the baby's internal mental state, by allowing her to distance herself from the observed feeling, relying more on a reflection of what might be going on for the child. This allows the child to see a mirror of him/herself, providing the infant with better coping strategies and resilience when facing distress (Fonagy et al., 2011).

Opposite mechanisms were found to develop in the case of insecure attachment or in parents with a history of trauma, whereby lower levels of oxytocin were measured, impeding on the mother's capacity to mentalize (Fries, Ziegler, Kurian, Jacoris, & Pollak, 2005; Heim et al., 2008; De Rosnay & Harris, 2002). In these instances, lower levels of oxytocin related to lower levels of attunement and mentalization in response to the child's distress, rendering the child vulnerable to mental states. Inner experiences therefore tend to become enacted rather than experienced and contained, leading to the development of an alien-self (Buchheim et al., 2009; Fonagy et al., 2011; 2012; Strathearn et al., 2009). The development of an alien-self, alongside biological vulnerabilities, lead to failures in mentalizing capacities in this child, increasing the risk for later psychopathology (Luyten & Fonagy, 2015).

Discussing the underlying biological and neurological mechanisms of mentalization, Fonagy (2008) found that it involves the frontocortical system, responsible for mental representations, and the mirror neuron system, focusing on the immediate response. In addition, it requires impulse control management, a capacity developing across childhood, thus supporting the view of a continuum of development from immature to mature (Fonagy, 2008; Jurist & Meehan, 2009). Furthermore, Luyten and Fonagy (2015) took on the task of explaining the neurobiology of the four dimensions of mentalization, emphasizing that different but interrelated neural networks in the brain governing each polarity.

The first polarity relates to differentiating between the affective and cognitive components of mentalizing. The former relates to empathy, more automatic and embodied, and is based on the activation of the ventromedial prefrontal cortex. The latter, more controlled is governed by the prefrontal cortex (Fonagy et al., 2002; Jurist, 2010; Luyten & Fonagy, 2015). This distinction between the two systems is paralleled in the other three polarities.

Studies have identified core neural networks activated for both polarities of the second dimension, self- versus other- focused mentalization (Frith & Frith, 2006; Lieberman, 2007; Van Overwalle, 2009); however, two different mechanisms are at play. One system governs the empathic response one has towards the other, based on a shared representation of a situation (Ripoll et al., 2013). This is based on automatic processes, activating mirror neurons and visceromotor centers for the understanding of affect (Lombardo et al., 2010). This capacity emerges in early development as it is based on a lower unconscious function, that of mimicking of emotions. The second system, mental state attribution (Ripoll et al., 2013), is more abstract, forming during adolescence and based on interpersonal relationships (Luyten & Fonagy, 2015). It involves areas of the brain responsible for perspective-taking and more cognitive tasks, such as imagining the other person's mind (Gweon, Dodell-Feder, Bedny, & Saxe, 2012).

Looking at the third dimension, internal versus external based, the former, focusing on making sense of one's internal states, has been found to be governed by areas of the brain responsible for active reflection, mainly the medial frontoparietal network, whereas the latter, based on the interpretation of facial expressions, activates the lateral fronto-temporoparietal network and temporal poles, responsible for less controlled processes. This has been explained from a developmental perspective whereby, early in life, infants tend to rely on external features of the caregiver and parents rely on the child's behaviors, in helping them make sense of their mental states and internal worlds, thus explaining the automaticity of this function.

With regards to the fourth polarity, implicit mentalizing has been described as a default position, serving an evolutionary purpose for survival, in that we automatically "read the mind of others" in order to know how to respond. This echoes Kovacs et al. (2010) showing that areas of the brain responsible for basic implicit mentalization develop around 18 months of age, relying primarily on sensory information, whereas the ones tapping into more explicit features of mentalization develop later, between the ages of 4 and 6 years, based on linguistic processing (see Baillargeon, Scott, & He, 2010; Beeghly & Cicchetti, 1994; Gweon et al., 2012; Kovacs et al., 2010 for a more detailed discussion of the neurological development of these capacities in childhood). The more controlled aspects of mentalizing are seen as crucial in correcting some of the interpretations made by automatic mentalizing, when facing stressful times or in response to threats. The former provides more accurate interpretations of mental states, based on a balanced analysis of perspectives, involving areas of the brain governing effortful control (Luyten & Fonagy, 2015). The authors theorized that psychopathology can be conceived in terms of temporary deficits in explicit mentalizing, explaining the different neurological underpinnings of the two extremes of this polarity.

So what could trigger these deficits in explicit/controlled mentalizing? Luyten and Fonagy (2015) discussed two main triggers: threat situations, triggering the fight or flight responses, but also the attachment system. Therefore, it was argued that one's attachment history plays a role in regulating the level of arousal necessary for the switch from automatic to controlled mentalizing to occur (Fonagy & Luyten, 2009; Luyten & Fonagy, 2015). This is important as it has informed trauma-based research; Mayes (2000) explained that the threshold for the automatic/ controlled switch decreases after exposure to traumatic events, mainly due to a more sensitive activation of the amygdala in response to stress and fear.

This understanding of the neurobiological basis of mentalizing has been used to inform mentalization based therapy, which will be detailed later. First, let's explore how the developmental task of mentalizing, coined as

Reflective Functioning within attachment relationships, develops in infancy and childhood within the context of the attachment relationships with primary caregivers.

Chapter 3

REFLECTIVE FUNCTIONING AND THE DEVELOPMENT OF THE SELF

Pia Tohme and Martin Kolev

THE DEVELOPMENT OF THE CONCEPT OF REFLECTIVE FUNCTIONING

Fonagy and his colleagues developed their own conceptualization of mentalization in an attempt to explain the intergenerational transmission of attachment. The concept was developed as part of the London Parent-Child Project, specifically looking at meta-cognition monitoring and coherence of attachment narratives of mothers and fathers (Fonagy et al., 1991). Reflective functioning (RF) was operationalized as the manifestation in speech of mentalization, specifically in the context of attachment-related interactions and narratives.

They defined maternal RF as the mother's ability to attribute feelings, thoughts and desires underlying her own and her baby's behaviors as well as to hold her baby and his mental states in her mind in a non-defensive way. This is theorized to facilitate his discovery of his internal world through the mother's representation of it (Fonagy et al., 1991; Fonagy, 2008; Slade,

2005). This idea echoes Winnicott's notion of the infant looking into the mother's eyes to see and validate him/herself as an intentional being with his own thoughts, beliefs and feelings (Steele & Steele, 2005; Winnicott, 1956). Extending the theory of the *unconscious* mind, Fonagy and Target (1997) stated that RF concerns "knowledge of the nature of experiences which give rise to certain beliefs and emotions, of likely behaviors given knowledge of beliefs and desires, of the expectable transactional relationship between beliefs and emotions, and of feelings and beliefs characteristic of particular developmental phases or relationships" (p. 680).

As part of the London Parent-Child Project, the RF scale was elaborated through a careful study of adult attachment narratives, focusing on the presence or failure to appropriately use mental state language when giving an account of parents' history with attachment figures (Fonagy et al., 1991; Steele & Steele, 2005). RF was conceptualized as determined based on 4 main types of mental functioning: 1) awareness of the nature of mental states, 2) explicit effort to tease out mental states underlying behaviors, 3) recognition of the developmental aspects of mental states, and 4) recognition of mental states in relation to the interviewer (Fonagy et al., 1991; 1997; Slade, 2005).

Fonagy et al. (1991) were the first to conclude that parents who scored high on RF tended to have a secure attachment with their own parents and that RF assessed prenatally could predict attachment security of the child at the age of 1. They explained that parents classified as secure were more likely to describe relationships in terms of mental states, therefore providing the infant with an environment promoting the development of a secure attachment relationship. In contrast, parents with a history of trauma and insecure attachment tended to rely on defensive strategies, such as splitting or denial, which limit their capacities to reflect on mental states. In line with previous findings (Main et al., 1985), it was found that these parents were less coherent when reporting narratives of their own attachment history and scored lower on RF. The authors theorized that this was likely to constrain their providing of an adequate psychological environment promoting the development of a secure attachment in the infant, as the offspring is not seen for who he/she is but as the parents' projections and distortions. This was hypothesized to partly explain the concordance of attachment between parent and offspring (Choi-Kain & Gunderson, 2008; Fonagy et al., 1991; 1998; Slade, 2004).

RF was conceptualized as comprising a self-reflective component as well as an interpersonal one (Fonagy et al., 1991; Fonagy et al., 2007). Grienenberger, Kelly, and Slade (2005) emphasized that the hallmark of reflectiveness is not only to recognize mental states in the self and other but also to be able to link and be aware of the interaction between mental states underlying self and others' behaviors. Slade (2005) made the distinction between adult RF based on attachment narratives between the parent and their own parents (Fonagy et al., 1991) and RF based on attachment narratives about the ongoing developing relationship between the parent and the child (Slade et al., 2005), the latter presenting a more direct measurement of the mechanism underlying the intergenerational transmission of attachment.

Slade et al. (2005) were the first to report that the mother's ability to reflect about her relationship with her own child was linked to both her attachment status with her parents and the children's attachment classifications. They explained that mothers who were classified as having a secure attachment with their parents were more likely to score high on RF, followed by mothers who were classified as dismissing or preoccupied, and mothers with a disorganized classification scoring lowest on RF. This shows that parents who are able to describe childhood experience coherently are more likely to ascribe mental states underlying their children's behaviors. Most importantly, RF was shown to be a central mechanism in the intergenerational transmission of attachment as research concluded that the relationship between adult and child attachment disappeared when RF was controlled for (Slade et al., 2005).

THE ROLE OF MENTALIZATION IN THE DEVELOPMENT OF THE SELF

Previous sections elaborated on the role of parental mentalizing in promoting the development of a secure parent-infant attachment relationship. However, what makes it crucial in the development of the self? RF has been argued to bring together two theories (Katznelson, 2014). First, the social biofeedback theory (Gergely & Watson, 1996) focusing on parental marked mirroring of the child's affect, leading to internalizing representations of the self, promoting self-regulation. Second, the psychic reality theory, discussing three modes of representations which evolve towards an integrated mode of mentalizing (Fonagy & Target, 1997). The latter described 1) *psychic equivalence* as a mode equating internal and external worlds, 2) *pretend mode*, during which the child can perceive separateness between the internal world and external reality, with the two being strictly apart and 3) *teleological mode* through which mental states cannot be internally represented and are based on or acted out in the external world (these three modes with be expanded on later through adolescent case illustrations).

There is a general agreement within the psychoanalytic field that self-organization starts with the infant's realization of his separateness and independence from the mother, that he/she is an independent and autonomous being, with a separate mind, thoughts, feelings and wishes. This initially creates frustration as needs cannot be satisfied instantly (Winnicott, 1962). With repeated interaction and synchronicity between mother and child, the latter slowly learns to control his impulses and manage his frustration (Bowlby, 1969). However, at this stage, the child is still in pre-mentalizing modes, as he/she cannot yet represent the caregiver's thoughts internally, despite starting to form stable internal working models of the parent, expecting and anticipating their responses (Fonagy & Target, 1997).

Around the age of 1, the child starts developing an awareness and a curiosity towards the other's mental states, instead of simple mirroring (Target & Fonagy, 1996). In line with classic psychoanalytic theories, Fonagy and Target (1997) illustrated this premise with an example of the child managing anxiety. Initially, anxiety is experienced by the child as a confusing mix of physiological and emotional uncomfortable feelings which he/she cannot yet explain (psychic equivalence). The mother in this case not only mirrors the child's feelings but gives it meaning by showing him/her she knows what that feeling is. She has therefore provided the child with a representational map, providing organization to the self and the other, driving him/her towards more mentalizing modes (Fonagy & Target, 1997).

This bridge between physical and emotional is only possible through sensitive parenting, which teaches the child to wonder about the underlying affective states explaining physical experiences (Slade, 2004). The child is therefore seen as a mental agent, "which is perceived by the child and used in the elaboration of teleological models, and then in the development of a core sense of mental selfhood" (Fonagy & Target, 1997, p. 690). In other words, the child now understands that his behaviors have an effect on the external environment and is able to modify it in his favor (Allen et al., 2008; Freeman, 2016).

Until the age of 3 or 4, the child is moving towards a state of pretense, during which he/she is attempting to build the bridge between internal and external experiences, as things may feel "too real", despite the tendency to want these two worlds to match (Freeman, 2016; Target & Fonagy, 1996). This starts to change with the development of pretend play, through which a tear occurs between these two worlds, with the child's assumption that external reality is unrelated to internal experiences (pretend mode). Within a healthy developmental environment, the child is then able to integrate these two modes and develop early stages of mentalizing, using the caregiver's elaborative mirroring (Fonagy & Target, 1997; 2005). According to Stern (1985), thoughts are perceived as real by the child, and can only be represented mentally through their recognition by more mature minds. Self-agency therefore develops through the acknowledgment of the child's intentional stance.

THE EFFECT OF FAILURES IN MENTALIZING

So far, we have discussed the importance and development of mentalizing capacities. In sum, Fonagy, Target, Steele and Steele (1998) presented at least 5 areas of relevance: 1) RF renders behavior easier to predict, 2) RF promotes and maintains attachment security, 3) RF facilitates the distinction between appearance and reality, 4) RF improves communication, and 5) RF helps making meaningful connections between internal and external worlds.

However, what is the effect of failures in mentalizing? Fonagy and Target (1997) suggested that within the context of a secure mother-infant relationship, the mother has the capacity to mirror the child's anxiety in a way that the child can understand what he/she is feeling, thus mapping within his/her self-representations. Similar mechanisms have been suggested to take place in instances of insecure attachment, in which case failed mirroring occurs due to a discrepancy between the infant's feelings and their representations by the mother (Fonagy & Target, 1997).

Dismissing parents tend to reject their children's signals in stressful situations because they awaken unresolved past trauma; thus, the parents provide inappropriate mirroring of affect to the child, which creates an insecure-avoidant response in them. Preoccupied parents tend to be more focused on their own feelings with current issues or past attachment figures, and are therefore unable to attend to the infant's needs, or respond excessively in order to compensate for previous negative experiences, which leads to an insecure-resistant child (Van Ijzendoorn, 1995). It was suggested that disorganized children tend to have an alien sense of self, and that the attachment relationship between the child and the caregiver revolves around externalizing parts of this alien self onto the attachment figure rather than working on internalizing a capacity of containment (Fonagy et al., 2007). This is in line with Slade's hypothesis that non-reflective parents tend to mirror a faulty image to the child, colored by their own insecure/disorganized attachment. This in turn leads to the construction of a wrong sense of self in the child, who is not seen for whom he is (Fonagy et al., 1991; Slade, 2005).

Within the framework of developmental psychopathology, Sharp et al. (2006) developed the accuracy of parental mentalizing paradigm. The task consisted of asking 7- to 11-year-old children to attribute thoughts to peers based on imaginary distressing scenarios, and then invite mothers to guess their children's responses. Looking at the concordance and discordance between the children's actual thoughts on stressful scenarios and the mothers' attributions about the child's thoughts, the researchers concluded that low maternal accuracy, thus low maternal RF capacities, promoted ineffective mentalizing in the child. The authors put forward that parents with distorted RF skills were unable to contain the children's negative and stressful affect because their reactions to the child depended on inferences they made based on unresolved issues within their past. These, in turn, were projected onto their children, leading to low social-cognitive reasoning skills and psychological adjustment (Grienenberger et al., 2005; Sharp et al., 2006). Steele and Steele (2005) suggested that some parents with low RF capacities lacked the ability to access the roots of behaviors and showed signs of isolation and disavowal of affect and a lack of interest in understanding one's own and the child's behaviors.

Based on the consistent findings highlighting the importance of the development of mentalizing and RF capacities throughout the lifespan, Fonagy and Bateman (2004) proposed its application within a therapeutic setting. The development and application of Mentalization Based Treatment (MBT) will be discussed next, tapping into the effects of these interventions on neural networks in the brain.

Chapter 4

MENTALIZATION BASED TREATMENT (MBT)

Pia Tohme and Martin Kolev

THE DEVELOPMENT OF MBT AND ITS PROCESS

Arguing the importance of integrating mentalizing within psychotherapeutic settings could be understood from two somewhat different perspectives: some argued that most psychiatric patients show evidence of an inability to understand the mind (Vanheule et al., 2011), whereas others presented the argument of dysfunctional mentalizing, arguing the development of this capacity, however putting it to use for unlawful means (Allen et al., 2008; Bateman & Fonagy, 2008). Here, it is important to note that mentalizing is not always used in order to promote a healthy understanding of others' minds. For instance, it has been shown that people diagnosed with Antisocial Personality Disorder are able to use the cognitive aspect of mentalizing, thinking about how the other might feel, aiming to manipulate him/her, thus failing to account for compassion, part of the affective component of mentalization (Allen, 2003; Fonagy & Levinson, 2004). Similarly, Allen (2003) explained how sadists use their mentalizing skills to humiliate or frighten others. In these instances, MBT is seen as useful in enhancing mentalizing about people and the mind rather than mentalizing objects. The goal of mentalization based treatments would therefore include a focus on helping the patient acknowledge the connection and bridge the gap between the body/physical reality and the mind/underlying mental states (Fonagy et al., 2011).

Noteworthy that the concept of mentalizing was partly developed based on Moran's work with borderline patients, highlighting a rejection of thinking about mental states, as a symptom of this psychopathology (Fonagy, 1991). The application of mentalization within the clinical field was therefore based on patients diagnosed with Borderline Personality Disorder (BPD), manualized as Mentalization-Based Treatment (MBT; Bateman & Fonagy, 2004). MBT for borderline patients is founded on the idea that stressful situations tend to lead to a failure in mentalizing (Bateman & Fonagy, 2014). This is based on a hyperactivation of attachment systems as a way to defend against stress and anxiety, which in turn, leads to inappropriate and intense attachment to others on the one hand, and an inhibition of the neural networks responsible for judging the level of trust we have of others, on the other (Bateman & Fonagy, 2013; Daubney & Bateman, 2015). However, when the BPD patient realizes that his/her needs are not met, strategies are reversed leading to extreme acting out colored by hostile and dismissive behaviors (Bateman & Fonagy, 2013).

Fonagy and Allison (2014) explained that the types of mentalizing identified in borderline patients resemble those of young children making sense of their internal world, namely psychic equivalence. In other words, in BPD, one has a distorted sense of intersubjectivity, based on the faulty assumptions that what is going on internally is an accurate reflection of the physical reality, with a failure to acknowledge potential alternative explanations for motives and behaviors. Another pre-mentalizing mode of thinking at play in BPD is pretend mode, in which it is impossible to imagine a bridge between internal and external realities. Finally, the teleological pre-mentalizing mode of thinking implies that one sets a judgement about the intent of others based on the concrete physical consequences of their actions as they occur in reality, with no other explanation conceivable. The use of these pre-mentalizing modes of thinking is likely to lead to an over-arousal of attachment systems, rendering it more difficult to re-establish balance in the self (Bateman & Fonagy, 2004; Daubney & Bateman, 2015).

Fonagy and Target (2005) explained that the first step towards enhancing reflective capacities would be to encourage one to observe affective mental states underlying behaviors. Within a therapeutic setting, the therapist helps the patient make the link between internal and external experience, introducing a curiosity about one's own and the other's experience (Allen, 2003). The rapport, or therapeutic alliance, is seen as mimicking a secure attachment relationship (Skarderud, 2007). The goal of MBT is set to improve patients' mentalizing capacities focusing on the here-and-now, through promoting mentalization and working through instances of failures in mentalization apparent through the transference.

This involves the person becoming more curious about his/her own and the other's mental states, beliefs and thoughts underlying behaviors, as well as discussing any discrepancy felt with regards to external experiences and their internal affective expression. This has been shown to favor the development of a stable view of the self and to maintain optimal levels of arousal within attachment relationships. In other words, the work within an MBT framework primarily occurs within the transference relationship, as the therapist makes use of these feelings to clarify to the patient what might be going on, encouraging him/her to acquire mentalizing capacities and stabilize emotional expression (Bateman & Fonagy, 2010; Fonagy & Target, 1998). MBT aims to foster in the patient a curiosity and an ability to consider alternative solutions and different perspectives to understanding one situation, namely the activity of mentalizing (Allen, 2003; 2006).

In the quality manual for MBT, Bateman, Bates and Hutsebaut (2014) presented seven main competencies and skills of the MBT therapist:

1. *Not-knowing, genuine and inquisitive therapist stance*: the therapist should model an authentic curiosity about the patient's internal world and mental states, focusing on a collaborative exploration, while being aware of the limits of one's knowledge of others.
2. *Support and empathy*: the therapist should provide empathic

responses to the patient's narrative and acknowledge, when appropriate, attempts at mentalization on his/her behalf.

3. *Clarification*: the therapist should check in with the patient to ensure a proper understanding of the narrative in an effort to make links between actions and feelings.
4. *Exploration*: the therapist should support the patient's curiosity about mental states, helping him/her overcome instances of non-mentalizing.
5. *Challenge*: the therapist should encourage the patient to see a different perspective.
6. *Affect focus*: the therapist should help the patient think and elaborate on mental states and affective processes.
7. *Relationship*: the therapist should make use of the here-and-now transference relationship in promoting mentalizing and exploring feelings or topics impeding these capacities, in order to focus on repairing the therapeutic relationship, serving as a model to other interpersonal relationships.

The Neurobiological Basis of MBT

As previously detailed, the main aim of MBT is to improve mentalizing capacities, leading to emotion regulation and a more balanced sense of self. Fonagy and Bateman (2006) contended to explain the neuroscientific basis of change in MBT, focusing on two psychological processes, the reward system and deficits in relatedness.

Investigating dysregulation in the reward system, Fonagy and Bateman (2006) explained that BPD patients struggle with impulsive behaviors and tend to favor short-term immediate safety and support. In fact, studies conducted in laboratories with BPD patients have identified that this population presents more difficulties in terms of money management, similar to people suffering from lesions in the orbitofrontal context, governing reward anticipation (Bazaris et al., 2002). Cardinal et al. (2001) had previously found that people diagnosed with BPD have trouble with delayed gratification, as they are more likely to choose options with short rather than long delays, despite the possibility of receiving a bigger reward if they waited. Research pinpointed to the activation of the nucleus accumbens and dopamine systems in this decision-making.

These findings are in line with results of studies with rodents, having identified similar brain areas implicated in bonding and attachment, labelling these behaviors as addictive (Insel, 2003; Insel & Young, 2000). Similar patterns of behaviors were identified in human mothers asked to look at pictures of their own versus others' children in different emotional states. Findings suggested that the orbitofrontal cortex plays a unique role in attachment behaviors, mainly the positive mood in relation to one's own offspring only (Nitschke et al., 2004; Strathearn & McClure, 2002). In sum, the neural circuits and the mesocorticolimbic system responsible for mediating reward, delaying gratification and addiction, also activated in attachment-based situations, were hypothesized to be dysfunctional in patients suffering from BPD (Fonagy & Bateman, 2006).

The second area of deficit in BPD is relatedness, as these patients tend to oscillate between idealizing others inappropriately, followed by a fast disillusionment if needs are not met, perceived as abandonment (Bateman & Fonagy, 2013; Fonagy & Bateman, 2006). This has been found to be associated with pre-mentalizing modes of thinking and difficulties differentiating between the self and others' mental states. This translates as a decreased activation of the medial prefrontal cortex and temporal lobes involved in tasks that have been described as challenging to BPD patients, such as assessing social trustworthiness (Winston, Strange, O'Doherty, & Dolan, 2002), interpreting the meaning of facial expressions (Critchley et al., 2000), making moral judgments (Greene & Haidt, 2002), and performing tasks that entail attending to one's own emotions (Fonagy & Bateman, 2006; Gusnard, Akbudak, Shulman, & Raichle, 2001).

Having described the two main areas of deficits in BPD, Fonagy and Bateman (2006) go on to explain the role of the MBT therapist in affecting these two systems, namely the role of the patient-therapist relationship through which the latter elicits discussions about past and present attachment relationships, as well as emotionally-laden situations, allowing the patient to mentalize within the therapeutic space. The patient is therefore encouraged towards countering the normal pattern of interpersonal relating, colored by a hyperactivation of attachment system, by, instead, being supported to stay in a mentalizing stance, based on patterns close to the ones used by people with a secure attachment relationship. However, this is only possible if this exposure is done gradually, taking into account the patient's level of readiness and insight, in order to avoid the patient being flooded with unbearable anxiety (Fonagy & Bateman, 2006). In sum, "the specific advantage of MBT in this process may be its focus on the simultaneous activation of the attachment system and encouragement of development of psychological processes that are normally inhibited as a result" (Fonagy & Bateman, 2006, p. 425).

MBT has been since adapted to different populations, such as parents (Cooper & Redfern, 2016; Sadler, Slade, & Mayes, 2006; Slade, 2007), adolescents (Bevington, Fuggle, Fonagy, Target, & Asen, 2013; Rossouw, 2015; Rossouw & Fonagy, 2012), children (Midgley, Ensink, Lindqvist, Malberg, & Muller, 2017; Midgley & Vroura, 2012; Sharp et al., 2006) and families (Asen & Fonagy, 2012), as well as towards treatment interventions for psychological disorders such as depression (Fischer-Kern et al., 2008; Taubner, Kessler, Buchheim, Kacheke, & Staun, 2011), trauma (Allen, 2003; Allen & Miga, 2010), eating disorders (Skarderud, 2007), somatic disorders (Luyten, van Houdenhove, Lemma, Target, & Fonagy, 2012) and antisocial personality disorder (McGauley, Yakeley, Williams, & Bateman, 2011) among others.

However, it is important to highlight that, despite the basic principles and guidelines remaining the same across all modalities, it is crucial to consider specific aspects of the population at hand and refining the structure and process of MBT to be efficient with each targeted population. The following section will focus on adolescent development in an attempt to highlight the crucial changes during this stage of development, leading to special considerations in MBT for adolescents (MBT-A).

Chapter 5

MAJOR CHANGES OCCURRING IN ADOLESCENCE

Pia Tohme and Martin Kolev

PSYCHOANALYTIC THEORIES

Psychoanalytic theories argued that, in adolescence, "one of the most significant, but also one of the most painful, psychical achievements of the pubertal period is completed: detachment from parental authority" (Freud, 1905, p. 227). The adolescent should now exclude parents as potential love-objects, despite this being the simplest course of action given that they constitute the people he/she has known and loved since childhood. In fact, the now more mature ego puts a barrier against incest, forbidding previously acceptable object-choices (Freud, 1905).

Anna Freud extended her father's ideas about the struggles of adolescence. She characterized this stage of development as a period of turbulent sexuality, during which the psychic balance achieved during latency is disturbed, due to the influx of libido caused by sexual maturity (Freud, A., 1958). According to the developmental line *From Dependency to Emotional Self-Reliance and Adult Object Relations*, she specified that, during adolescence, there is a great effort around loosening the infantile ties to parents and a defense against pre-genitality, occurring before puberty, in order to move to a more acceptable choice of love object (Freud, A., 1958). This process entails an emotional struggle and a mourning process, in order for the libido to detach itself from the oedipal strivings towards parents and cathect new appropriate heterosexual objects. This development is only made possible by some narcissistic withdrawal to fill the gap made by the lack of an appropriate external object to cathect (Freud, A., 1958).

Jacobson (1961) suggested, paradoxically, that the ego reinstates past positions in order to relinquish past attachments and gain the optimal and acceptable instinctual freedom needed to find new ways of instinctual discharge to build on adult relationships. This genuine development is based on a revision and selective acceptance or rejection as well as a flexible mastery of substitutions of aims and representations of self, other and relationships. This process starts with the loosening of ties to parents and reasserting one's role within the family structure, as earlier psychic formations now have a subordinate role (Deutsch, 1944; Schafer, 1973). It is argued that these changes allow the child to move towards a conscious experience of the self as a mature adult (Gourevitch, 1980).

Blos (1967), based on classic psychoanalytic ideas and Mahler's (1963) concept of separation-individuation, compared adolescence to infancy, as they both have in common firstly, the urgency in psychical change in order for development to happen, and secondly, a heightened vulnerability. He emphasized that one is therefore more prone to psychopathology if the individuation process fails during these developmental stages (Blos, 1967). What Mahler (1963) described in infancy as hatching from the symbiotic relationship with the mother, becomes in adolescence, according to Blos, a pulling away from parental infantile dependencies (Steinwand, 1984). Adolescence was thus conceptualized as a period of second individuation, characterized by a regressive pull, during which the adolescent learns to manage the tension between primitivization/ differentiation and regressive/progressive positions (Blos, 1967; Tyson & Tyson, 1990). These changes are thought to render the ego and the self very fragile, leading to feelings of apprehension and uncertainty (Ammaniti, 1988). Feelings of anxiety and aggression also prevail as the adolescent finds ways to channel these feelings in a meaningful way, in the service of the goals and aspirations of the ego's changing needs, in order to healthily go through this stage and attain independence from infantile ties (Gourevitch, 1980).

All abovementioned theories focused on the concept of regression which has been defined in psychoanalysis and modern psychology as "a reversion to earlier forms in the development of thought, of object relationships or of the structure of behavior" (Laplanche & Pontalis, 1973, p. 386). Blos (1979) described resistance against regression, which is related to the defense mechanisms presented by Anna Freud. He added that some of the forms it can take are negativism, oppositionalism or indifference as ways of asserting oneself as an independent and individual persona. It can also take the form of a turn towards the outside world in a reversed attitude, giving the adolescent an illusionary and apparent victory towards parents.

According to Blos, the more dependent a child was on his parents during the early years, the more distance he will take from parents during adolescence to assert independence (Blos, 1967). This is in line with the idea that one of the most typical defensive strategies is the externalization of conflicts, as a way to avoid the threat of passivity or submissiveness to parents and the dissolution felt by the renewed dependency on them. Most importantly, externalizing conflicts allows the adolescent not to think about those painful feelings or acknowledge the separation and individuation taking place (Ammaniti, 1988). For the restructuring of the ego to be possible, it was hypothesized that the adolescent will need, after temporary regression and use of defenses, to be able to renounce infantile ties and cope with the stress it entails, as well as accept his independence from parents (Jacobson, 1961). Given this context, it can be argued that classical psychoanalytic theories view adolescence as a time of emotional turbulence during which strong drives confront a weak ego.

THE ROLE OF PARENTS AND PEERS IN IDENTITY FORMATION

Adolescence is a period during which parents cannot play the same role as they did in childhood; they are de-idealized and dependency on them gradually decreases as they are not seen as the authority figure anymore (Allen, 2008; Blos, 1967). These developments lead to the superego's loss of some of its significance given that it was based on identification with, and internalization of parents. The adolescent now has the choice to select the aspects of parents to identify with and modify the ideal self-image into a more realistic one. This is first perceived as an emptiness and a painful alienation, thus enticing a turn to non-judgmental peers as a source of support during the resolution of detachment to infantile ties. The internal conflict is then externalized as conflict with parents, as the adolescent comes to terms with selecting and internalizing some of the parental standards, in order to integrate them with past identifications, representations and ideals, as well as the moral standards taken from peers.

These developments constitute a further step towards the psychic re-structuralization, consolidating the adult personality. Once this is achieved, the superego is fully internalized, independent from parental authority, and functions under ego supremacy (Tyson & Tyson, 1990). Along the same lines, Laufer (1964) described the adolescents' contemporaries and friends as potential auxiliary-egos or auxiliary-superegos as they may temporarily assist the ego in defensive acts. In fact, during adolescence, the ego needs external allies to overcome feelings of guilt and anxiety, which are found in friends and contemporaries with whom he identifies (Brandt, 1977; Laufer, 1964).

A closely related concept is that of the ego-ideal which was first described by Freud in 1914 in "On Narcissism" and subsequently defined by Laplanche and Pontalis (1973) as "an agency of the personality resulting from the coming together of narcissism (idealization of the ego) and identification with parents with their substitutes or with collective ideals" (p. 144). In adolescence, the ego-ideal constitutes the ideal the person wants to be, including identifications with parents, peers or idols, reflecting the meeting of inner wishes or the avoidance of conflicts (Laufer, 1964). Giovacchini (1973) emphasized the importance of the adolescent's involvement in the socio-cultural milieu, outside the familial boundaries, as a reaction to the upsurge of self-involvement. These transient identifications play a major role during adolescence as they also include a social component without which the adolescent loses a guiding principle giving life direction, continuity and meaning, making this agency crucial in the process of detaching the self from childhood dependencies on parents (Blos, 1967).

Erikson (1956) focused on another crucial achievement of adolescence, identity formation. He argued that self-identity is reached when all the experiences of temporary self-diffusion, caused by the regression occurring during that period, are successfully contained and integrated. He also argued that the process of identity formation is constantly evolving in order to integrate biological changes, identifications, defenses and libidinal needs. Some role repudiation is needed in order to delimit identity through experimenting with roles before reaching one's identity (Erikson & Erikson, 1997). According to Erikson (1970), the basic patterns of identity emerge from two main areas, repudiation and choice of childhood identifications and, the way society and the community accept and recognize the adolescent. He also differentiated between two unhealthy resolutions of this stage. On the one hand, role diffusion, where the adolescent's ego seeks to expand the boundaries of the self for a sense of wider identity, gaining certainty and conviction. On the other hand, role confusion, defined as "states in which there is an impoverishment and a dissipation of emotional, cognitive, and moral gains" (Erikson, 1970, p. 15).

Marcia (1966) expanded on Erikson's theory of identity and presented the identity status paradigm, focusing on two main dimensions. The first parallels Erikson's identity crisis, stating that there should be an exploration of identities within various domains, and the second, that of commitment, defined as the selection of alternatives and the engagement in activities that lead to the consolidation and implementation of the adolescent's choices. Based on this theory, Marcia classified adolescent identity into four identity statuses based on the amount of exploration and commitment. Identity diffusion occurs when the adolescent is low on both exploration and commitment, foreclosure describes an individual who is low on exploration but high on commitment, the opposite combination leads to a moratorium identity status, and finally a person who is high in both domains would be classified as achieved (Kroger, 2006; Marcia, 1966).

COGNITIVE AND EMOTIONAL DEVELOPMENT AND THE ADOLESCENT BRAIN

Studies focusing on the neurochemical and cellular changes occurring as a function of age, are based on animals. Despite most mammals and some rodents exhibiting increased social interactions during "adolescence" (Primus & Kellogg, 1989), the simple extrapolation of findings has been questioned with one social phenomenon in mind: the human adolescent, solely, undergoes the psychological stress of this developmental stage. This tends to occur alongside an increased seeking for new sensations and risk-taking behaviors (Adriani, Chiarotti, & Laviola, 1998).

So what explains these behaviors? During early puberty, new neural connections develop in the limbic and prefrontal region, followed by pruning in later adolescence. Only as much as half of the synapses developed during childhood will be preserved during adolescence (Crews, He, & Hodge, 2007). A neuropsychological report by Casey, Jones, and Hare (2008) outlined the brain dynamics of the adolescent, emphasizing three main developments.

First, they described dendritic pruning in the amygdala, nucleus accumbens and prefrontal cortex, as well as continual growth in the density of the fibers connecting the amygdala and prefrontal cortex from adolescence into early adulthood. These findings are consistent with the idea that nucleus accumbens matures earlier than the prefrontal cortex.

Second, structurally, the white matter in the adolescent brain is more pronounced in the frontal area and the grey matter development continues, reaching peaks in the frontal and parietal lobes at age 12, temporal lobe until age 16 and occipital lobe until age 20.

Third, consistent with the cellular and structural changes, Casey et al. (2008) also discussed alterations in neurotransmission in the cortical and subcortical regions as outlined above. Crucial for the communication between nucleus accumbens, amygdala and prefrontal cortex is the dopamine secretion, which holds the balance between excitation and inhibition of impulses. Noteworthy is the overproduction of white matter and peak levels of dopamine in early adolescence, constituting a neurobiological explanation of the impulsive and often "unbalanced" adolescent behavior.

Emotional Developments

Emotional responses in adolescents tend to be more extreme as they become triggered by abstract ideas from the past or anticipated future responses, as well as the development of an understanding and ability to predict other people's feelings and to consider the differential impact one situation can have on the self or others (Rosenblum & Lewis, 2006). This can be explained by the far-reaching neurobiological and psychological organizations taking place, as well as developments in the adolescent brain.

Authors have argued that the adolescent's newly found independence can often be overwhelming for adolescents, which has led some researchers to characterize this period as one of "storm and stress" (Somerville et al., 2010). However, between the 1960s and 1970s, these ideas came under attack as being based on clinical samples and not necessarily generalizable to the wider population. Studies conducted in schools and community samples led researchers to agree that perturbations do occur, with the parent-adolescent relationship going through substantial transformations, but not to the extent of alienation and rebellion (Douvan & Adelson, 1966; Kandel & Lesser, 1972). Comparing this development period to one of "storm and stress" has been found to be an exception rather than a rule (Phares, Vicky, Renk, Duhig, Fields, & Sly, 2009). Only 5% to 15% of adolescents experienced these extreme situations, which origins were usually based prior to adolescence (Smetana, Campione-Barr, & Metzger, 2006; Steinberg, 2001).

Two areas of growth are most prominent during this developmental period. Firstly, many of the changes in the brain happen in areas that control the regulation of behavior and emotions, as well as the evaluation of risk and reward. Secondly, changes in the arousal system brought about by the physical maturation preceding the development of appropriate regulatory systems. According to Steinberg (2005), these changes could be compared to "a situation in which one is starting an engine without yet having a skilled driver behind the wheel" (p. 70). This is in line with Somerville et al. (2010) who suggested that risk-taking behaviors should not be considered as a deficit in cognitive ability to comprehend and anticipate the consequences of one's actions, given that, cognitively, adolescents do have the necessary capacities to foresee negative and dangerous outcomes. They added that, instead, the environmental context, including peers, as well as emotional states, take over cognitive capacities and cloud the adolescent's judgments. Moreover, brain developments during this developmental stage can be compared to brain maturation in the first five years of life given the extensive and crucial changes that occur (Storelder & Ploegmakers-Burg, 2010).

Another major development during adolescence is affect regulation, involving controlling one's feelings based on societal rules, in order to achieve goals (Dahl, 2004). Scholars differentiated between cold cognitions, one's ability to affect regulate under low arousal conditions, and hot cognitions, including situations of high emotional stress and arousal, the latter more problematic in the case of adolescents (Dahl, 2004). In other words, "in emotionally salient situations, subcortical systems will win out (accelerator) over control systems (brakes) given their maturity relative to the prefrontal control system" (Casey, Jones & Somerville, 2011, p. 22). Contrary to popular belief, these changes can be mainly understood in terms of neurological changes and neurotransmitter functions rather than hormonal changes (Moretti & Peled, 2004).

To conclude, by early adolescence, the capacity to control emotions starts to emerge and is efficient in most situations. It is not before middle to late adolescence that young people can express and accept inner conflicts, explaining them in terms of feeling states, thus reflecting the development of a capacity to hold contradicting feelings towards one situation (Harris & Gross, 1988; Saarni, 1984). However, scholars converge in noting that adolescents' cognitive skills develop slower than emotional capacities, shedding light on the increase of reckless and impulsive behaviors during this stage of development (Dahl, 2004; Steinberg, 2004).

Cognitive Developments

With regards to cognitive developments, Steinberg (2005) explained that, throughout adolescence, brain areas controlling executive functioning, including long-term planning, self-regulation and self-evaluation develop. It was suggested that this allows the adolescent to integrate information more autonomously, think in abstract terms, make decisions and predict emotions more accurately (Blakemore, 2008; Storelder & Ploegmakers- Burg, 2010). The cognitive brain, as explained by Storelder and Ploegmakers-Burg (2010) includes the hippocampus which is the center of learning and explicit memory, the dorsolateral prefrontal cortex which is the base of executive functioning, including long-term planning, self- regulation and self-evaluation, and the medial prefrontal cortex which holds mentalizing capacities. This explains adolescents' improvements in reasoning and information processing, as well as hypothetical thinking, which carries on developing through adolescence (Steinberg, 2005).

Another crucial aspect governed and influenced by the changes in the prefrontal cortex is the development of higher cognitive capacities such as awareness of the self but also of the other, or the theory of mind (Blakemore & Choudhury, 2006; Frith & Frith, 2003; Ochsner & Gross, 2004). These have been found to be affected by both physical and hormonal changes in adolescence. Importantly, social cognition is also deeply rooted in environmental influences, affecting the level of synaptic pruning depending on the amount and quality of social interactions (Blakemore & Choudhury, 2006). This, in turn, promotes or demotes the development of perspective-taking and mentalizing capacities, crucial to social communication and problem-solving.

One widely accepted change in adolescence is an increase in risk-taking behavior, involving difficulties in impulse control. As discussed by Mischel, Schoda, and Rodriguez (1989), impulse control and its manifestation in behavior can be observed and assessed through the construct of delayed gratification. In their seminal experiment (Mischel & Ebbesen, 1970; 1972), a preschool child was seated in a room with the reward of his choice in front of him (a cookie or other) after which the experimenter left the room. The child was instructed that if he/she does not eat the treat during that time, he/she will be given a bigger reward, therefore given the choice between an immediate but small reward and a bigger reward provided after a short period of time. Results revealed an increased ability to postpone and control impulses with age, in the presence of highly salient incentives (Mischel et al., 1989). More recently and based on these findings, Eigsti et al. (2006) concluded that differences in impulse control appear to remain throughout adolescence and young adulthood.

Other studies focused on investigating adolescents' ability to differentiate between good and bad decisions. In an fMRI study, Baird, Fugelsang, and Bennett (2005) presented participants with scenarios, asking them to rate whether they thought they represented a good or bad idea, tracking neural mechanisms to determine differences between adolescent and adult brains. They concluded that adolescents spent more time judging "bad idea" scenarios, with areas of the brain relating to reasoning capacities being more active, showing they put more effort than adults in decision-making with regards to risky behaviors. By the end of adolescence, synaptic pruning, the strengthening of connections between prefrontal and subcortical regions and the increased speed of neural information in the prefrontal cortex, allow for the fine-tuning of cognitive control, more specifically relating to judgement-making and differentiating appropriate from inappropriate actions (Blakemore & Choudhury, 2006; Casey et al., 2011).

Similarly, Steinberg (2005, 2008) proposed the social reward model to explain adolescent risk-taking behavior and decision-making, highlighting that the prefrontal cortex is not yet efficient enough to regulate emotional responses. Accordingly, the adolescent is more likely to be influenced by external factors such as peers and social feedback, thus being more prone to peer pressure. Investigating age differences in risk-taking behavior, Steinberg and his colleagues found that adolescents tend to engage in the same amount of risky situations as adults when alone; however, risky behaviors almost doubled in adolescents when they thought they were being watched by their peers (Gardner & Steinberg, 2005).

Social Developments

Early scholars discussing adolescence compared this stage to any other stage of development (Collins, 2003; Collins & Laursen, 2004). However, contemporary scholars, echoing psychoanalytic theories, have come forward with a more complex understanding of this transition phase; in parallel to the physical, emotional and cognitive changes occurring, the adolescent faces another challenge, identity formation in the context of peer and romantic relationships (Collins & Laursen, 2004; Moretti & Peled, 2004). Peers come to play a crucial part in trying out different social roles in the process of identity formation. Similarly, dating becomes at the forefront of adolescents' worries (Collins & Laursen, 2004; Furman & Buhrmester, 1992).

The social brain is "the complex network of areas that enable us to recognize others and evaluate their mental states (intentions, desires and beliefs), feelings, enduring dispositions and actions" (Blakemore, 2008, p. 267). It comprises 1) the medial prefrontal cortex, necessary for the understanding of one's own and others' intentions, 2) the anterior cingulate cortex, the inferior frontal gyrus, the superior temporal sulcus, which govern complex movements, 3) the amygdala, responsible for the processing of emotional facial expressions, and 4) the anterior insula, which is involved in the development of thinking about intentions, perspective-taking, irony, and mentalizing capacities, defined as the capacity to understand and interpret thoughts and feelings of the self and others in term of mental states

(Blakemore, 2008; Fonagy et al., 2002, Sebastian et al., 2010; Steinberg, 2005; Strorelder & Ploegmakers-Burg, 2010).

Research done with children between 9 and 14 years of age showed that the medial prefrontal cortex and the left inferior frontal gyrus are more active, reflecting the brain's intention to resolve the discrepancy between the literal, expected meaning of a remark and the intended meaning during ironic interactions. These same areas are activated in adults in mentalizing tasks (Amodio & Frith, 2006; Gilbert et al., 2006). Strorelder and Ploegmakers-Burg (2010) explained that the mentalizing capacities continue to develop through adolescence, as during this period, the young adult is more sensitive to reading facial expressions, signifying acceptance or rejection. Furthermore, the maturation of the medial prefrontal system, area of the brain involved in this process, carries on developing even after the young adult reaches twenty-one years of age (see mete-analysis by Blakemore, 2008).

Crone and Dahl (2012) differentiated between social-cognitive development, including mentalizing capacities, and social-affective development encompassing social acceptance and rejection, related to the understanding of social feedback from peers. One study compared brain activity in adolescents and adults, highlighting that social rejection was more distress-provoking in adolescents than in adults, as shown by an increased activation of the ventral anterior cingulate cortex (ACC). However, these results were buffered by the presence of a strong support group, thus speculating that having strong peer relationships plays a protective role against rejection in adolescence (Gunther Moor, van Leijenhorst, Rombouts, Crone, & Van der Molenref, 2010; Masten et al., 2009; Masten, Telzer, Fuligni, Lieberman, & Eisenberger, 2012).

Masten et al. (2009) and Sebastian et al. (2010) devised a series of video games with simple turn-taking tasks that required two players, one being the participant and the other supposedly another participant in an adjacent room; however, the second player is in fact a computer programmed to stop playing after several turns. Affective consequences of this exclusion were then measured specifically looking at mood changes and affect regulation in female adolescents when facing perceived rejection from high interest peers, highlighting a significant lowering in mood and increase in anxiety levels compared to adults (Sebastian et al., 2010). Findings showed a relationship between social competence and emotional distress with adolescent scoring high on rejection sensitivity more likely to feel emotional distress evidenced by an activation of the subgenual anterior cingulate cortex (Masten et al., 2009). In sum, social inclusion and recognition is considered as a major factor in reward seeking and heightening vulnerability to risky and reckless behavior with the objective of social acceptance and inclusion.

Investigating parents' responses to these changes, researchers have found that parents' validating and empathic responses to their adolescents' feelings were linked to better emotion regulation skills and conversely, dismissing and hostile responses from parents were related to later internalizing and externalizing behaviors (Davidov & Grusec 2006; Gottman, Katz, & Hooven, 1997; Klimes-Dougan & Zeman, 2007; Morris, Silk, Steinberg, Myers & Robinson, 2007, O'Neal & Magai, 2005). Therefore, it can be argued that parents' flexibility and modifying of their affective responses in order to match the changes occurring in their adolescents could lead to a better adjustment and might be apparent at the heart of the parent-adolescent relationship.

These brain developments are facilitated within a secure attachment relationship with parents, providing a template for peer and romantic relationships (Armsden & Greenberg, 1987; Collins & Laursen, 2004; Feldman, Gowen, & Fisher, 1998). Parent and peer relationships play distinct but complementary roles in adolescence with parents providing support and guidance and peers serving affiliative functions (Kobak & Ferenz-Gillies, 1995; Kobak et al., 1991). Peers have also been described as providing a "secure base" to the adolescent moving towards the dating world and romantic relationships (Connolly et al., 2000; Connolly, Craig, Goldberg, & Pepler, 1999; Furman et al., 2002).

Are Adolescents More at Risk of Developing Mental Health Problems?

By the end of adolescence, the disengagement from internalized objects should be complete in order for the ego to acquire stable representations of the self and other. The previously detailed processes lead to a more prominent role of the ego-ideal and a lessening influence and rigidity of the superego. These structural changes establish constancy in self-esteem and mood, independent of external sources of satisfaction (Blos, 1967). The fear of parental influence due to the regression to oedipal strivings and fantasies, the persisting wishes to remain close to parents and the idealized views of parents is now replaced by a genuine emancipation based on revisions and flexible mastery of patterns of behaviors (Schafer, 1973). The adolescent now takes responsibility for who he/she is and for the choices made as he/she attains a sense of identity, maintaining a sense of sameness and continuity which matches that of one's meaning to others (Ollech & McCarthy, 1997).

From a parallel, albeit theoretically different perspective, other scholars have presented adolescence as a stage of "strength and resilience" (p. 3), as the adolescent goes through a series of maturational physical, cognitive and emotional developments (Dahl, 2004). This capacity for resilience can be understood in terms of neural plasticity during this stage of development, facilitating learning of new strategies and problem-solving skills (Blakemore & Choudhury, 2006).

Bringing it all together, it can be posited that adolescence is characterized by a qualitative shift in the way one perceives and makes sense of him/herself and the world. Various cognitive, social and emotional developments allow him/her to be more self-aware and reflective, holding in mind different, at times contradicting, perspectives in a strategic manner (Blakemore & Choudhury, 2006). In other words, adolescents are able to reflect rather than react. In sum, Dahl (2004) argued that, despite adolescence starting with the physical changes brought upon by puberty, it does not end in the biological domain solely; instead, adolescence ends with the "attainment of adult roles and responsibilities" (p. 9), based on the interaction between physical, cognitive, emotional and social developments.

All of these changes have been hypothesized to increase the adolescent's risk of developing mental health problems. Some scholars argued that the critical developmental tasks of adolescence, mainly achieving autonomy from parents, while maintaining a positive relationship with them and developing new friendships, could lead to some struggles or strains, increasing the chances of the development of depressive symptomatology (Allen et al., 2006; Kobak & Ferenz-Gillies, 1995; Kobak, Sudler, & Gamble, 1991). Allen et al., (2006) identified withdrawal, anger and dependent behaviors as interrelated factors within peer relationships increasing depressive symptoms. Others discussed peer contagion, as a facilitator of delinquency or aggression in adolescence, a process not under full conscious awareness, but rather, motivated by the adolescent's need for belonging and fitting in (Dishion & Tipsord, 2011). Sawyer et al. (2012) discussed findings in relation to adolescent health, focusing on their heightened sensitivity to peers as affecting their potential for substance use and misuse and its timing (Patton et al., 2002; Trucco, Colder, & Wieczorek, 2011), with the same applying to sexual experimentation (Parent, Teilmann, Juul, Skakkebaek, Toppari, & Bouguignon, 2003). In their model, Ary and colleagues concluded that the family environment, including parental conflict, low parental monitoring or low family involvement, increased the chances of the development of risky behaviors in adolescence (Ary, Duncan, Biglan, Metzler, Noell, & Smolkowski, 1999).

Bongers, Koot, Ende, and Verhulst (2003) took on the task of identifying developmental trajectories of parent-reported problems in adolescence in an attempt to further understand their onset at this stage of development. Some relevant findings relate to an increase in scores on withdrawal and anxious-depressive problems for both boys and girls, differing in onset depending on the onset of maturational pubertal changes for each. With regards to externalizing problems, aggression, delinquent and antisocial behaviors were all found to decrease in late adolescence. These findings provide initial developmental trajectories of some problematic behaviors in adolescents along which some deviances occur. Neuroscientific studies have supported these conclusions, speculating that neurological developments occurring during this critical period could partly explain the onset of these internalizing and externalizing behaviors in adolescence (Nelson et al., 2005; Paus, Keshavan, & Giedd, 2008).

Mentalization-based treatment has been put forward as one approach integrating findings from the neuropsychological field in order to promote resilience and healthy resolution of this transition stage (Fonagy, Rossouw, Sharp, Bateman, Allison, & Farrar, 2014; Rossouw, 2015; Rossouw & Fonagy, 2012).

Chapter 6

IMPORTANCE AND DEVELOPMENT OF MENTALIZATION BASED TREATMENT FOR ADOLESCENTS (MBT-A)

Pia Tohme and Martin Kolev

What the Mentalization Based Treatment for Adolescents (MBT-A) structure holds as a core belief is, first of all, to provide safety for the adolescent, but also to the family as a whole. This is rendered more complicated if the therapist is solely relying on "snapshots" of the adolescent's life to explain some phenomenon; this would only mean that the approach failed before it even started. Rather, the mentalizing therapist favors a more holistic view of the adolescent individual, considering the abovementioned roles of peer pressure, risk-taking, reward-seeking and emotional reactivity. The MBT-A therapist thus takes on the task of incorporating these elements in a welcoming developmental framework, holding the process while respecting autonomy, even if autonomy is not palpable at the start of treatment. In other words, the MBT-A structure can be seen as a holding environment within which containment of the thinking process can take place, as opposed to acting out.

SPECIAL CONSIDERATIONS IN MBT-A

As detailed above, interpersonal relationships play a crucial role in the adolescent's world. However, the development of the limbic system governing emotions precedes that of the prefrontal cortex responsible for more rational thinking and making sense of others' intentions and motives (Steinberg, 2005). These developments render the adolescent more prone to interpersonal stress, thus hindering mentalization and making it more likely to act out impulsively in an attempt to regulate internal states (Rossouw, 2015). It is therefore argued that the MBT-A model plays a role in strengthening the adolescent's mentalizing capacities when dealing with stress, thus promoting self-control and emotion regulation instead of concrete acting out (Bleiberg, 2013; Fonagy, 1998; Rossouw, 2015).

However, it is important to adapt the basic MBT model taking into account the specific developmental changes and influences of adolescence, as detailed in Chapter 5. In terms of the MBT-A structure this translates in weekly meetings with the adolescent, but also monthly meetings with the adolescent and his/her parents together. The individual MBT-A sessions with the adolescent have no particular structure, but similar to the MBT model presented in Chapter 4, focus on mentalizing the ongoing patient-therapist relationship and other interpersonal issues the adolescent might be facing, focusing on thinking about the mental states evoked. The goal of the family MBT sessions is to enable all family members to be curious about what might be going on for the other and how these influence family dynamics (Bleiberg, 2013; Rossouw, 2015). MBT-A also includes an increased focus on affect and emotion regulation within the therapeutic space, as a way to help the adolescent pick up and contain the increased impulsivity characterizing this developmental stage, as these hinder mentalization (Rossouw & Fonagy, 2012). In sum, MBT-A is presented as a model focusing on affect rather than cognitions, forming accurate representations of the self, others and relationships in order to restore mentalizing (Rossouw, 2015).

Pre-Mentalizing Modes of Thinking in Adolescence

As previously described, Bateman & Fonagy (2014) coined the term teleological mode as a phenomenon through which the individual bases the interpretation of internal mental states on concrete facts. This makes it difficult to consider alternative explanations and perceive different perspectives, putting one in a non-mentalizing state.

It can be argued that, given the biological changes occurring in adolescence and the slow-developing prefrontal cortex, it becomes more difficult, if not impossible, for the adolescent, to mentalize without the help of a more mature mind. For instance, it is not uncommon to hear the following from a 12-year old girl, after her best friend told her he now has a girlfriend "See, I told you he won't love me anymore! He has a girlfriend now, he won't even hug me to say hi anymore… he is always talking about her! He doesn't even listen to me when I talk! It is all about her, her her! And he hugs her! He will just forget about me now".

This example of teleological mode of thinking reflects the anxiety and paranoid thoughts this girl is having based on behaviors and acts, or lack of her best friend hugging her. These overwhelming thoughts are simply translated as feeling abandoned and forgotten, without the possibility of seeing alternatives. She is unable to hold contradicting feelings in her mind, being happy for her best friend while at the same time accepting changes in her relationship with him. This, of course, remains hard even past adolescence; however, the role of MBT-A interventions would be to *Pause and Rewind*, in order to open the lines of communication with this girl, considering a different explanation, or discussing how it might feel for her to be rejected or abandoned, without getting overwhelmed and shutting down mentalization. For instance, the therapist might say "Hold on a second, could we just pause and rewind. I feel this is really upsetting you. Too much change too quickly, I want to try to understand what might be going on for you." This reflection could help restore emotion regulation and decrease impulsivity, allowing the adolescent to test out other explanations.

As Rossouw (2015) summarized, "misperceptions and assumptions about the mental states and intentions of others often lead to extremely strong feelings that can tip into impulsive behavior, even before the feelings register consciously" (p. 181). Therefore, rewinding and reviewing interpersonal conflicts between adolescents and parents or peers, help them in becoming more aware of signs of emotion dysregulations, allowing them to better self-control and mentalize, becoming less vulnerable to anxieties and self-esteem collapse.

Another example would be that of a 16-year old girl, throwing a jealousy fit because her boyfriend did not instantly answer her text messages, despite having read them. "He was online! I saw him! And he read my messages and didn't answer. I know he is talking to that girl. I knew she liked him and would try to make a move on him. He is asking her out, I know it!"

This girl is relying on psychic equivalence, equating internal and external worlds, as she is certain of her boyfriend's disloyalty without any proof of it being right. In this case, the MBT-A therapist acknowledges how difficult and anger-provoking this might feel to the adolescent who is perceiving abandonment and wonders, with the patient, whether there are other alternatives. "I cannot know how this must have felt for you but you sound pretty angry with him. I wonder whether we can think together what might be going on there… Imagine your best friend talking, what do you think she would have told you?" Sometimes, getting the adolescent to think of what one of the peers would have said tends to facilitate the kick start of mentalizing, as it allows them to take a step back from their dysregulated self and look at the interpersonal distress from a different perspective, a hallmark of mentalizing.

Another important consideration highlighted by Rossouw (2015) is that flexibility is required with adolescents in establishing the rapport and trust with them. This could, at times, be represented by changing the setting of the session, either sitting on the floor or taking a walk, depending on what might facilitate the adolescent's discussion of sensitive topics. Teleological interventions, reflecting the therapist's flexibility in finding the appropriate tools to connect with the adolescent, are considered useful as they allow the adolescent to visualize what might be going on in the mind. This concrete interpretation plays the role of a transitional object, facilitating discussions about more abstracts internal mental states.

For instance, this 14-year old girl states waking up in the middle of the night because she sees and hears shapes and shadows; however, she fears working with a therapist as "getting better would mean making those voices and thoughts go away, but where would that leave me, who will I be? I am no one if those voices are gone." Within the MBT-A model, the therapist first acknowledges how scary this must feel for her, providing an empathic validation of her feelings, "This sounds terrifying, not knowing who you are and how you might change. I wonder whether we can try to put this down on paper to make sure I really understand. Sometimes I feel that trying to draw these helps clarify our abstract thoughts."

The therapist continues "It looks to me that, if we try and take these thoughts away, it might create some gaps in who you are, a feeling of emptiness maybe… this is really scary."

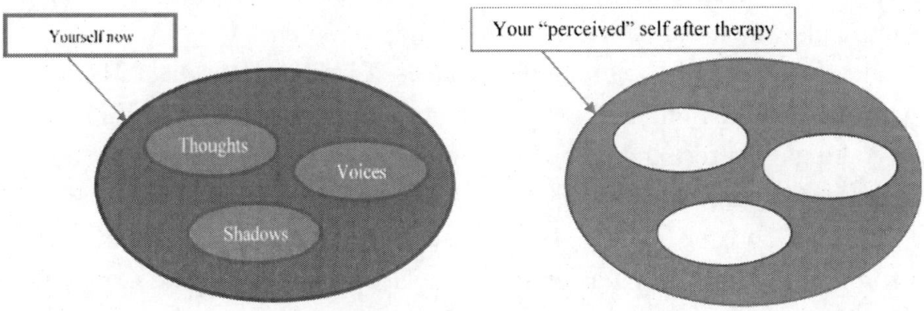

This teleological representation can help the adolescent visualize the small objectives of attempting to fill each "gap" with her own identity, instead of feeling that the emptiness might take over the self and that she might lose herself in it. The therapist serves as a model and a source of support, encouraging mental state talk about and exploration of overwhelming and sensitive topics.

Finally, consider this 13-year old boy who, according to his teachers, was being inappropriate in answering them back and has no boundaries. He had already seen the school counselor who listened to his recounting of the injustice he felt, not being heard by his teachers who would blame him for mistakes he had not done. He perceived this as disrespect on their part which, to him, justified his disrespect towards them (pre-mentalizing). The school counselor attempted to use a teleological representation by asking him whether he could be more like a leaf, which bends with the wind in order not to break to which he answered "why would I want to be a leaf if I can be a big strong mountain?" Noteworthy that this adolescent was being bullied by his friends for being overweight. This example not only highlights the pre-mentalizing defensive modes of thinking, but also sheds light on the importance of considering the role of the body during adolescence.

The Adolescent's Relationship with the Body

A challenge of adolescence is developing one's identity, including the acceptance of one's body image. Despite not being the main focus of this chapter, it is of interest to introduce Lemma's conceptualization of the role of the body in adolescence in order to better grasp the complexity of this developmental stage. In her book, Lemma (2014) discusses the four assumptions which she views as pillars of her thinking.

First, she states that "embodiment shapes the mind" (Lemma, 2014, p.2) suggesting that the mind and the body cannot be understood as separate, as they continually affect and guide each other. Second, she discusses the body ego, emphasizing the importance of bodily emotions and sensations in both the therapist and the patient. Lemma sees the interaction between both interpretations of sensation as crucial in a successful therapeutic approach. Third, she theorizes on seeing and being seen as playing a role in the development of the self, be it within the parent-child relationship or the therapist-patient relationship. Fourth, she discusses the role of our perception of the maternal body in better understanding our own feelings and phantasies.

Lemma also discusses the role of cyberspace and technology in allowing to modify or create an illusion of embodiment which does not necessarily coincide with the real self, thus creating a disruption with reality (Lemma, 2015). This separateness allows for the body to be denied, thus creating what Zizek (2004) coined as "pseudopresence", which he argued decreases the effects of an emotional experience or even avoiding it. Lemma (2015) further argued that, in cyberspace, internal and external worlds are one and the same, one in which psychic equivalence therefore dominates. This feeling of omnipotence could lead the adolescent to having less established boundaries, given that what is imagined could be real, as one can create an online identity (avatar), reflecting the ideal self (somatic flexibility). Lemma (2015) argues for the importance of the adolescent's awareness of the effect of the incongruence between the ideal virtual self and the real offline self, which she refers to inauthentic embodiment.

Given the role social media and technology currently plays in identity formation in adolescence, we can therefore argue that the "cyberclosness" could therefore induce a decrease in emotional experiences, which could in turn lead to pseudomentalizing.

MBT-A as a Brain-Based Treatment

Over the last several decades, neuroscience has, in fact, focused a great deal of attention on the biology of experience, elucidating the ways in which life experiences affect the structure and functions of the brain. Every time we see, hear, smell, or touch something, learn a new fact, or have a new experience, new proteins are synthesized and neural pathways communicate the new information to multiple brain regions.

Neuroscience is the field exploring brain connections underlying the various overt behaviors or changes occurring throughout development. It can be argued that neuroscience also helps elucidate brain changes resulting from psychotherapeutic interventions. Gorman (2018) summarized how Cognitive-Behavioral Therapy (CBT) and psychoanalysis relate to changes in specific brain areas. CBT has been found to be successful in treating anxiety disorders and phobias, based on a decreased activation of the amygdala, partly responsible for signaling fear (Furmark, Tillfors, Marteinsdottir, Fischer, Pissiota, Langstrom, & Fredrikson, 2002; Bijsterbosch et al., 2018). The CBT mechanisms are based on asserting rational thoughts, overcoming cognitive distortions and irrational fears triggered by an overstimulation of the amygdala. From a psychoanalytic standpoint, Gorman and Roose (2011) proposed that fears and phobias can be conceptualized as reconsolidation of fear memories. Through psychoanalytic interpretations, the therapist helps with the reactivation of that memory and the creation of a less fearful reconsolidation of it.

Healthy emotion regulation seems to require recognition of the emotional significance of perceived stimuli, appreciation for the need for regulation and selection/implementation of appropriate strategies (Sheppes, Suri, & Gross, 2015). This perspective was elaborated upon by Kesek (2010) who added that social cognitive skills, more specifically perspective taking, are a crucial factor for emotional regulation. It is therefore easier for the adolescent to consider a different perspective to a current situation if he/she is emotionally regulated; here, we argue that emotion regulation relies on perspective taking as a major contributing social cognitive skill. Considering the various changes occurring in the adolescent brain, it can be suggested that adolescents may be particularly vulnerable to emotional dysregulation, although it is worth noting that, as with adults, it can be difficult to distinguish whether these behaviors result from poor regulation, increased affective responses or both. MBT-A, as a perspective taking paradigm, relies on both and postulates that adolescence is in fact a critical phase for the development of adaptive emotional regulation strategies. It also suggests that adolescence does have a heightened potential for learning and flexibility (Casey et al., 2008; Steinberg, 2005).

According to Wekerle (2010) there exists a window of opportunity that can be targeted through the implementation of adaptive emotion regulation strategies, leading to long-term positive results regarding mental health. In fact, given the numerous changes occurring during this stage of development involving synaptic pruning and reorganization, it can be argued that adolescence is a sensitive period, particularly vulnerable to environmental influences, both positive and negative (Blakemore & Choudhury, 2006; Steinberg, 2005). It is also characterized as a period of increased problem with behavior and emotion control, sometimes leading to resort to psychological support or therapeutic interventions. For instance, the incidence of mood disorders and affect-related disorders increases during this developmental stage, partly due to increased emotional reactivity or sensitivity in adolescence (Casey et al., 2008; Pine et al., 2001; Steinberg, 2004).

Developed as an approach for treating Borderline Personality Disorder (BPD), MBT holds the belief that these individuals are vulnerable to a loss of mentalizing capacity due to their affective arousal, becoming overwhelming and too intense. The same can be argued in terms of adolescence. Studies in the past two decades focused on the mythical sense of invincibility that often perceived in adolescence. Garderner and Steinberg (2006) theorized that adolescents can think and reason adequately in social situations but the more mature limbic system and affective "fuel" overwhelms the prefrontal cortex's capacity as a regulatory and rationalizing buffer.

MRI studies in adults have shown that their prefrontal cortex activates when they think about other people's minds. Adolescents too are capable of such a complex ability; however, it seems more accessible if facilitated by a more mature mind within the context of a trusting and safe enough environment, encouraging an imaginative approach and curiosity to other peoples' minds. These findings are seen as a connecting bridge towards the process of mentalizing the other and a basis of MBT-A.

So, how would this look within a session? The MBT-A practitioner acknowledges the effort to explore the other's mind and supports it by evoking curiosity about a main mentalizing theme, "Could you let me know how it is to be you? Because I am not you and I don't know… but I would really like to understand…" In this short example, the MBT-A therapist is actively questioning, engaged in perspective taking, warmly containing in the therapeutic context and using his own self in the process. The goal is for the therapist to mentalize his own self and then to participate in mentalizing the adolescent in the room. This same process – sounding easier than it is – views emotional regulation as main entrance towards perspective taking. In other words, in order to mentalize, one has to be below "the heat of the moment" and above "the state of emotional stillness".

There are many strategies aiming at emotion regulation, one of the most prominent being the ability to focus on the specific moment of time at which regulatory processing has been brought to dysregulate. According to this model emotional generation and appropriate regulatory processes unfold in a particular sequence over time (Gross, 1998; 2014). This seems to correspond with what MBT-A views as exercising proper mentalizing over time in order to establish a more adaptive stance. The first two processes of the Gross model are situation selection and situation modification, both of which help with shaping the situation to which the individual will be exposed.

In sum, what the MBT-A structure holds as a core belief is, first of all, to provide safety for the adolescent, in the first instance, and then for the family as whole. The mentalizing therapist would take into account the abovementioned importance and role of peer pressure, risk-taking, reward-seeking and emotional reactivity and incorporate them within a safe and good-enough developmental trajectory, respecting the adolescent's autonomy. So, the MBT-A structure can be seen as a holding environment within which containment can take place with the MBT-A therapist actively questioning, engaging in perspective taking and warmly containing within the therapeutic context, using his own self in the process.

Chapter 7

CONTEMPORARY ACHIEVEMENTS OF NEUROSCIENCE

Vanya Matanova and Zlatomira Kostova

Neuroscience is an interdisciplinary field that cooperates with medicine, psychology, physics, chemistry, computer science, linguistics, mathematics, philosophy, physics, etc. Neuroscience has a direct connection to genetics, biochemistry, and physiology, and it studies the morphology and function of the central nervous system. The term neuroscience is regarded as a complete study of the nervous system, its molecular, cellular, structural, functional, evolutionary and medical aspects.

Brain processes became available for follow-up about 20 years ago, although neuro-technologies have been known for about 50 years. Both psychology and physiology study normal and disordered behavior as a result of a disorder of brain functioning. It is already known that nerve cells in different brain areas are not specialized for individual processes and functions. In the cortex, this specialization is related to the individual experience while in the more ancient structures - to the experience of the species. Behavior is influenced by all-brain systemic phenomena common to neurons in various structures.

Changes in brain functioning are difficult to detect and that is where translational research takes over. Thanks to it a change has been found in the brain at a cellular and molecular level, determined by past experience. In recent years, more and more research has been associated with the so-called translational medicine. Translational medicine aims at transferring (translating) fundamental scientific research to clinical practice. In turn, translational neuroscience seeks to integrate some basic *in vivo* brain activity studies in meeting the needs of patients who suffer from nervous system diseases. The research complex is based on the innovative Translational Validity Theory. According to it, there is a correspondence between clinical tests and brain activity measured by functional magnetic resonance imaging. Translational medicine is directly related to personalized medicine that targets a particular disease, taking into account the genetic, metabolic, etc. peculiarities of the individual.

In recent years, translational medicine has been trying to accelerate the application of biomedical research results in various medical fields. Medical and biological research focuses on the mechanisms of disease and the ways of influencing these mechanisms. These studies are the path to understanding the biology of certain diseases and the ways for their treatment. The main thing here is the process of translating the results from fundamental research into their practical application - from "the laboratory to the patient's bed". The current paradigm of research is based on generating hypotheses from different systems rather than from human beings.

The purpose of translational research is to join and combine scientific fields, resources, experience and techniques to improve prevention, diagnosis and therapy, i.e. the 'transformation' of scientific discoveries in laboratory conditions into potential treatments for various diseases (Cohrs, R. J., Martin, T., Ghahramani, P., Bidaut, L., Higgins, P. J., & Shahzad, A., 2015). "The problem of translation of assessment methods and data across neuroscience, phenomenology and clinical psychology has been addressed in the model of translational validation (Zachar, P., Stoyanov, D., Aragona, M., & Jablensky, A. (Eds.), 2015).

Recent discoveries have identified the so-called rosehip neurons, which make up about 10% of the first layer of the neocortex. It has been found out that rosehip neurons are characteristic only for the human brain and they connect with the pyramidal cells, which are two-thirds of all neurons in the cortex. The connection between rosehip and pyramidal neurons is still unclear, but researchers have found out that rosehip neurons act as inhibitory neurons, inhibiting the activity of other neurons. They have the potential to put brakes on the excitability of pyramidal neurons. (Trygve Bakken, 2012). But how this influences behavior is still unknown. This discovery raises questions about the methodology and translation of results.

Still, the attempts to introduce translational models in psychiatry, clinical psychology and psychotherapy in order to operationalize diagnostic information from different disciplines are of uncertain validity. There are two main objections related to mental disorders:

- the presence of qualia and
- multiple realizability.

Qualia are an argument related to phenomenological psychopathology and refer to the qualitative dimension of human experience. Multiple realizability refers to the fact that there are different brain parts involved in some psychopathological phenomena (Stoyanov, D., 2018). Perhaps this explains why some past psychic traumas leave traces and influence mental and social functioning and why often their negative effect remains despite the therapy. The methods of functional neuroimaging help to identify these changes. A number of studies show that our memory of past events is still poorly studied in neuroscience. The recollection of life events resembles more 're-building' than 'digging out.' Important life events change brain functioning. But the traces of the past can be so dramatically changed that their interpretation may end up based on entirely new constructions. The ability of the brain to restore a certain memory may be permanently deleted, but some new experiences may trigger the same activation as the previous event.

There is increasing evidence of the relationship between mind and body. Changes in mental states influence not only emotions but also genes. Under the influence of stress/distress, the hypothalamus transforms nerve impulses into neurohormonal ones and they are carriers of information for the body. This is a stimulus for the endocrine system to produce steroid hormones that reach the nuclei of different cells and cause gene modulation. These genes cause the formation of molecules that influence metabolism, growth, degree of activity, sexual functions and immune system responses both in health and disease.

The neurological concept of consciousness is related to the state of activity of the nervous system and the brain and it is the brain that changes qualitatively depending on the degree of activity - from high activity to its complete absence in coma or anesthesia. Different brain areas are involved in mental processes, but mental activity depends on the functioning of a group of neurons in the hypothalamus. If the functioning of the diencephalon is the basis of activity, then, perhaps, consciousness is located in that region? The tendency to divide etiological factors into mental and somatic ones will be sustainable as long as the soul-body dualism opposes any rational solution. We need a new formulation of this attitude based on objective realities.

Knowledge of brain functioning has been constantly evolving, which leads to periodic rethinking and updating of therapeutic approaches to various nervous system diseases and brain dysfunctions.

So far, there have been a few biological ways to check the effectiveness of different psychotherapeutic approaches. Functional tomography of the brain makes it possible to trace the dynamics of human behavior and the workings of the brain. If changes caused by psychotherapy are long term, it means that different forms of psychotherapy cause different structural changes in the brain, as do different forms of education. According to E. Kandel, in this way, it will be possible to study the psyche from top to bottom and to develop the program that Freud predicted.

MIRROR NEURONS AND MENTAL FUNCTIONING

There is convincing evidence for the active nature of mental processes and their immediate dependence on the individual's purposeful activity. These areas, which in classical neurophysiology are divided into motor and sensory areas, in fact are morphologically as well as functionally inextricably linked both to each other and to other brain structures. There is a system of neurons in the brain that encodes the individual's own actions and gives a functional meaning to the surrounding space. G. Rizzolatti called these neurons 'mirror neurons'; he thought they were "able to classify the incoming sensory information on the basis of the spectrum of potentially accessible actions, beyond the dependence on whether or not these actions will be accomplished" (Rizzolatti, G., Sinigaglia, C., 2008).

Mirror neurons have selectivity, they respond only to the observation of a particular action. One of their most important qualities is to react in a similar way both to own actions and to the ones observed. These properties are not considered in isolation but as a consequence of the functioning of a complex system of interacting cortical areas. They can use not only visual but also auditory information for the actions of others but do not have any particular function. Through the system of mirror neurons, the brain matches the image of the action with the action itself. They ensure in the brain the formation of representations of the action performed by someone else and its description in the terms of the observer's own motor act. According to G. Rizzolatti, the motor knowledge of our action is a necessary and sufficient condition for an immediate understanding of the actions of others. This mechanism is used for the realization of different functions depending on what aspect of the observed action is encoded, in what neuronal networks mirror neurons are involved, and how these networks are connected to other brain structures.

Verbal mirror neurons in the human brain play a particular role both in the realization of simple imitative motor acts and in social knowledge and behavior. The main specificity is that they get activated in one and the same way when performing certain actions and when observing these actions being performed by someone else. Studies of 'communicative' mirror neurons in monkeys showed that they respond not to actions with objects (transitive), as the other mirror neurons, but to actions without any objects at all (non-transitive) which are, however, important to communication (gestures, postures, lips positioning).

The encoding of non-transitive actions important for communication is more relevant to humans. These neurons can encode not only others' individual actions but also sequences of actions.

Analyzing these data G. Rizzolatti emphasizes that, despite the broader range of tasks these systems can perform, their main task is to 'understand the actions of others ... in the terms of purposeful movements,' i.e., 'the mirror mechanism fixes the intentional aspect of the action, which is common to the observer and the doer.' It is important to note that 'understanding the actions of others' in a given context does not imply any meaningful reflection but substantially depends on the 'action vocabulary' of the individual, which, especially in communicative actions, is type-specific, besides, it is also largely determined by individual experience. The special role of mirror neurons in social knowledge was also supported by a study of autistic individuals (Iacoboni, M., Dapretto, M., 2006).

The data about the work of mirror neurons of the frontal and parietal cortex ('frontal-parietal network') give reason to assume that the mirror network encodes exactly the purpose of the observed motor act, whose accomplishment is possible in different ways and with the support of various executive organs (Rizzolatti, Sinigaglia, 2010).

In transcranial magnetic stimulation, it has been proven that the mirror neuron system is capable of 'generalizing' with regard to the organs through which one and the same motor act can be realized. Thanks to these properties, mirror neurons are involved both in understanding the actions and intentions of others, and in understanding the causes of these intentions. They, however, are not the only mechanism used for understanding the whole variety of other people's behavior.

'Emotional' mirror neurons are located in the cortex between the frontal and temporal lobes of the brain. They respond in a similar way when experiencing certain emotions and when observing the face of another person. In this case, sensory information is encoded not in terms of action, as in “motor” mirror neurons, but in terms of emotional experiences. According to G. Rizzolatti, thanks to the activity of ‘emotional’ mirror neurons, an immediate understanding of others’ emotions is carried out in the same way as the actions of others are understood due to the activity of motor mirror neurons. This is the result of ‘splitting the emotional state of the other at the level of visceromotor reactions.’ Rizzolatti and Sinigaglia believe that this is fundamentally different from empathy. Our empathy depends not only on the recognition of the other’s pain but also on our relationship to them, our inner world, the level of our self-knowledge and many other factors (Sinigaglia and Rizzolatti, 2011). Complex forms of social behaviour, including empathy, cannot be reduced only to the activity of mirror neurons, but their role in brain mechanisms of social interaction is still valid.

The discovery of mirror neurons makes it possible to reformulate the principle of ideomotor compatibility - the common representative base should be seen not as abstract and amodal, but as a mechanism transforming visual information into motor acts (Iacoboni, M., Woods, R. P., Brass, M., Bekkering, H., Mazziotta, J. C., & Rizzolatti, G., 1999). In transitive actions, the effective interaction with the object is not a must, since the image of the action is sufficient for the activation of the mirror system. The mirror system and its selectivity contribute to the creation of a shared space of actions where each action and chain of actions - our own or others’ - are instantly registered and recognized without the need of explicit or voluntary cognitive operation. The mirror neuron system plays a crucial role in imitation processes, re-coding the observed actions into motor codes, thus making it possible to repeat them. Forms of imitation depend on the activation of certain cortical areas with mirror properties, which means that there is a mechanism behind them that compares the visual information received from observing the other’s motor act and the corresponding motor representations. Mirror neurons encode both transitive and non-transitive motor acts and track the temporal aspects of the observed action. The richness of motor repertoire, however, is not sufficient to develop the ability to learn. In other words, the mirror neuron system is a necessary but insufficient condition for imitation. This applies not only to the ability to learn by imitation but also to the ability to repeat an action, performed by another, that enters into our own motor repertoire. Imitation requires a system managed by mirror neurons that perform two functions - relaxation (facilitation) and suppression (inhibition). Sounds familiar, right?

The Frontal Cortex and Mental Functioning

Studies of brain frontal lobes, within the norm and with lesions, provide information on the identification of neuronal networks that have complex interactions with cortical and subcortical structures. The longer period of maturation of the frontal lobes in comparison to other brain structures puts them at risk of the effects of various pathogenic factors, which affects the mental and social functioning. The evidence from these studies helped for an integrative approach to psychotherapy and interventions in different conditions.

As far back as the 70s of the 20th century, A. Luria believed that the frontal lobes of the brain were the apparatus that ensured the formation of sustainable intentions that defined conscious behavior. The beginning of studies on the role of the frontal lobes in the organization of purposeful behavior was set with the observations of the consequences of their damage. The frontal lobes are complex in their construction and with numerous connections to cortical and subcortical structures. On one hand, they provide an opportunity for processing and integrating different affiliations and, on the other hand - different types of regulatory influence.

In cases of local damage to the frontal lobes, there are two groups of symptoms. The first one is associated with impairment of the mediobasal regions of the frontal lobes. There is a decrease in tone, rapid exhaustion and delayed reactions at the end of the implementation of the tasks. The emotional background is depressed, which is accompanied by depression, sadness, fear, and vegetative reactions; 'catastrophic reaction' syndrome is present, an 'end of the world' experience. In traumatic, tumor or vascular injuries of the mediobasal frontal lobes, there are disorders in the emotional-personal sphere in the form of an inadequate attitude towards oneself, towards one's own condition or illness, towards one's relatives and friends, non-criticism and lack of responsibility. Personal changes are especially pronounced in cases of damage of the orbital surfaces of the frontal lobes - disinhibition of nutritional and sexual attraction. The second group of symptoms is related to impairment of the lateral convex lobes. Symptoms include apathy, indifference, loss of interest in the surrounding world, adynamia and aspontaneity.

When the frontal lobes are injured, the forms of conscious activity and behavior that get predominantly damaged are those determined by motives mediated by speech activity. Behavior becomes unintentional, non-adaptive, the assessment of external influences changes as well as the coordination of sensory and motor components of action, the synthesis of target-oriented movements (A. Luria, 1969). Based on the research carried out, A. Luria united the brain structures responsible for the programming of movements and actions, for the regulation of ongoing active processes and for comparison of the effect of an action with the intent, into the Third functional block of the brain.

One of the main functions of the frontal lobes is motor behavior. The motor and premotor areas of the cortex and the oculomotor field are located there. All of them carry out the planning and implementation of voluntary movements - praxis. Broca's center, which is responsible for speech motor realization, is located in the lower part of the frontal curve (almost always in the left hemisphere). The prefrontal cortex plays a role in the formation of personal characteristics and emotional reactivity. Bilateral injuries of the frontal lobes are accompanied by attention deficit, decision-making difficulties, and anti-social behavior. Attention deficit leads to disorders of both short- and long-term memory. Injury of the prefrontal cortical lobe affects the ability to think as well as the ability to make deductions and predictions, the ability to draw hypotheses and the ability to suppress impulses. The characteristics of attention, flexibility of thinking and the ability to suppress inadequate reactions are impaired. Gradual slowdown in the pace of task execution is typical, which is defined as Dysexecutive syndrome.

A number of studies have been conducted in an attempt to determine the exact areas of the brain that lead to executive dysfunctions; this, in turn, has led to a huge amount of information which is often contradictory. A common hypothesis is that they are related to pathology in the prefrontal cerebral lobe (Clark C., 2002). Various studies have shown that executive dysfunctions are determined by the interaction between the frontal cortex and other regions such as the basal ganglia and the cerebellum (Giedd, J. N., Rapoport, J. L., 2010). Neuroimaging studies have shown that some functions are not localized in the prefrontal lobes (Carpenter P., just M., Reichle E., 2000). Studies with magnetic resonance have affirmed that there is a common network of regions in the frontal, parietal and occipital lobes, the thalamus and the cerebellum. This shows that executive functions are carried out by dynamic and flexible networks, which are characterized by integration and connectivity. The main scheme starts from the dorsolateral prefrontal cortex / orbitofrontal cortex, goes through the striatum and the thalamus and returns to the prefrontal cortex. The dominant opinion is that cognitive processes are provided by networks that encompass a multitude of cortical areas that are in close co-operation and overlapping functions (Carpenter, P. A., Just, M. A., Reichle, E. D., 2000).

Neuroimaging techniques have allowed us to specify the localization of executive functions in the brain regions. Among the cortical postcentral regions, the most significant role in securing executive functions is played by the parietal regions located around the intraparietal sulcus (*Sulcus intraparietalis*). In M. Posner's works (Posner, Petersen, 1990; Posner, Fan, 2008), these regions were regarded as key units in brain organization, securing the shifting of visual attention. It is well-known that both the frontal and the parietal regions of the cortex are activated in different situations that require the participation of selective attention mechanisms - this is what has produced the term 'frontoparietal attention system' (Ptak, 2012). The realization of executive functions is also carried out through interaction between the structures of the limbic system - the amygdala, the hippocampus, and the hypothalamus, together with the basal ganglia and the mediodorsal nucleus of the thalamus.

Even in cases of extensive bilateral lesions of the prefrontal lobe, perception, motor functions, and intelligence can be preserved; personality and behavior, however, change sharply. This is demonstrated in various real-life situations, where there is no control over behavior. In cases of frontal abulic syndrome, the initiative decreases as well as creativity and curiosity; indifference and apathy appear. In cases of frontal disinhibition syndrome there is impulsive behavior and inadequate views and judgments, difficulties in the use of past experience, manifested as a persistent repetition of one and the same inadequate acts without guilt. The prefrontal cortex, due to its extensive connections with other associative regions, plays an integrative, coordinating role with respect to the functional systems of the brain.

It is well-known that cognitive functioning is lateralized in the prefrontal and parietal cortex, the pre-motor, and the supplementary motor area. Part of the anterior cingulate cortex is activated in tasks that include making choices in conflicting information, detection of multiple goals and tasks related to the working memory (Banfield, Wyland, Macrae, 2004). The dorsal region is involved in the modulation of attention, executive functions and working memory and is activated during cognitive tasks (Bush, G., Luu, P., Posner, M. I., 2000). Though effective control is not required for all cognitive processes, it is activated under circumstances that require inhibition of the first possible response, error monitoring and correction, decision making and planning (Diamond, 2005).

Many scientists suggest that the level of creativity and intellectual work depends on how our brain is organized. The right frontal lobe, which is responsible for creative thinking, the processing of large volumes of information and the generalization of ideas, plays a crucial role in this respect. There is empirical evidence that artists deal better with non-specific tasks requiring imagination and allowing deviations from the ultimate goal, and much worse with specific, clear tasks.

Efferent connections of the prefrontal cortex and the amygdala are regarded as a possible substrate of descending control of emotional states (Banks, S. J., Eddy, K. T., Angstadt, M., Nathan, P. J., & Phan, K. L., 2007Sa; lzman, C. D., Fusi, S., 2010). According to Banks (2007). Cognitive control of negative emotions is ensured by the functional interaction of neuronal networks in the prefrontal cortex and the amygdala. The control over the external expression of emotions and somatic processes that accompany emotional states is carried out by direct and indirect connections of the prefrontal cortex and the hypothalamus. (Zikopoulos, B., Barbas, H. (2006).

When the information is updated, there is activation of the frontal and parietal cortical areas. These regions are activated in cases where it is necessary to switch from one cognitive task to another, regardless of the modality of the stimuli. This raises the question whether the frontal lobes are the only ones that ensure executive functions. Contemporary studies affirm that frontal lobes interact with different deep brain structures and other associative cortical areas.

The hippocampus also participates in the realization of executive functions. The hippocampus, together with the prefrontal cortex and other cortical and deep brain structures, is part of a functional neuronal ensemble that is formed during the period of retaining the information in working memory (Gazzaley, A., Rissman, J., D'esposito, M., 2004). When the load of working memory increases, the degree of activation of the hippocampus increases as well (Rissman, J., Wagner, A. D., 2012). The hippocampus rapidly combines individual elements of events into an integrated representation. Thus, the specific role of the medial prefrontal cortex in securing the organization of behavior in new, uncertain situations is accomplished by its interaction with the hippocampus in the process of integration of new information and past experience.

Executive functions are the control authority of the brain. They enable people to learn, adapt to the environment and function successfully and efficiently. They are crucial to the success at social, school and professional levels. Executive functions allow the individual to successfully coordinate and apply their cognitive abilities, skills and knowledge to achieve a given goal. The aspects of planning and organizing the activity are found to be impaired in case of damage to the prefrontal structures of the cortex. In these cases, the internal control of the activity is changed for an external one. Often the current situation begins to dominate and to affect behavior, which in Gestalt psychology is called 'field behavior.' Depending on the location of the lesion, a problem may occur related to the coordination of action components in time or difficulties with the cessation of action may arise, in which case perseverations occur (Goldberg, 2004). Activity disorders are accompanied by disorders in social functioning and in the structure of personality. Integrative processes and those related to reflexive awareness are particularly affected. An adequate assessment of the situation is possible, but there will be difficulties in decision-making or in case the decision has already been made – difficulties in its realization. In case of damage to the prefrontal lobes of the cortex, the nature of subjective perception of a situation is changed, the feeling of new emotions disappears and everything seems to be long known.

New data prove that the prefrontal cortex plays a role in personal characteristics and emotional reactivity. Dysfunction in this region adversely affects thinking, the ability to make deductions and predictions, the ability to draw hypotheses and the ability to suppress impulses. The characteristics of attention, flexibility of thinking and the ability to suppress inadequate reactions are impaired. Their development during psychotherapy is likely to affect the structure and functioning of the prefrontal cortex.

Analyses of the interhemispheric differences between left and right frontal lobes show that behavioral disorders in the form of impulsive actions and lack of self-criticism are more typical for the right lobe; whereas the left frontal lobe is more closely connected with intellectual processes and disorders of volitional regulation of mental activity. Highly differentiated emotional experiences associated with past experience disappear in case of injury in the left frontal lobe.

Modal non-specific disorders of attention are associated with damage to the upper sections of the brainstem and parts of the reticular formation. The frontal lobes, the thalamus, and the limbic system are engaged. Studies conducted at a neuronal level have shown that a significant place in the limbic system - i.e., the hippocampus, the amygdala and the caudate nucleus - is occupied by neurons that help erase old stimuli, thus ensuring a reaction to new signals and suppression of reactions to already common stimuli. When focusing on something, the middle part of the prefrontal cortex is activated. This is the so-called "frontal attention system". The involvement of other sections of the cortex is determined by the specificity of the stimuli and the nature of processing up-to-date information. The 'posterior attention system' includes the executive mechanism of the spatial selection of visual stimuli.

If mediobasal or temporal regions are damaged, volitional attention is affected, though non-volitional is increased, which leads to 'field behavior' - unusual, uncontrolled reactivity to all stimuli, determined by the poor functioning of the sensor filter. Modal-specific disorders of attention are manifested in a specific sphere and are described as a phenomenon of ignoring stimuli of a given modality. But this is not an agnostic manifestation, an intellectual deficit or misunderstanding of the instructions. Visual or auditory attention disorders are an initial sign of damage to the posterior lobes of the right hemisphere and may later evolve into agnosia - visual or auditory.

Memory is regarded as a complex functional system where the information entering brain structures is encoded. In order for the traces to remain a certain cortical tone is needed which is provided by the deep brain regions - the reticular formation, the thalamus and the limbic system. Injuries to the mediobasal regions of the frontal and temporal lobes lead to disorders of short-term semantic and conceptual memory. The right hemisphere is predominantly involved in immediate reproduction, while the left one is used for postponed reproduction. In case of left-side injuries, immediate reproduction is better preserved than in right-sided injuries.

Until recently, it was assumed that working memory was based on the ceaseless activity of neurons that process current incoming information. New research suggests that information may be stored in short-term memory not on the basis of increased neuronal activity but on the basis of short-term changes in synapse strength, which is associated with a change in calcium-ion concentration in the presynaptic terminal (short-term synaptic plasticity) (Cowan, N., 2017).

The Limbic System and Mental Functioning

The structures of the limbic system and its multiple connections with the hypothalamus and cortical structures lie at the core of emotions. In 1937, James Papez identified the structures of the limbic cortex as the cortical substrate of emotions. This is a ring of phylogenetically old cortex around the brainstem, parts of which are the amygdala and the hippocampus.

The temporal zones and the limbic system play a significant role in mental functioning. They are the center, which is empathic not only to emotions, passions, desires and spiritual experiences but also to memory and social connectedness. The temporal regions and the limbic system are responsible for encoding and storing events in long-term memory. With the help of the amygdala, the present and the past are integrated and fully perceived. In the case of decreased activity of the limbic system, a person is in an optimistic mood, and when the activity is increased - in a negative mood. The limbic system and medial part of the temporal lobe are a substrate thanks to which situations are valued emotionally. When they are in high activity, even neutral events are perceived negatively, which causes a certain behavior that hinders social interactions.

A pronounced relationship is established between the modality of emotions and the neurochemical processes in brain structures. Different biologically active substances - neuromediators - are mediators in the process of synaptic transmission of impulses and exert a certain influence on the emotional background. Excess or deficiency of noradrenaline, serotonin, gamma-amino acids, dopamine, adrenaline, endorphins, etc. give rise to different emotional states. Depending on different brain structures – the hypothalamus, amygdala, basal ganglia, limbic system - states such as fear, aggression, pleasure or panic are generated. Depending on the specific conditions and the physical state, one and the same mediator, hormone or bioactive substance can cause different experiences. Dopamine, for example, is associated with positive experiences and increased motor activity, and noradrenaline deficiency leads to depression.

Characteristics of emotional disorders in local brain injuries often carry a controversial and unclear meaning in the specialized literature. In case of injury of the frontal lobes, there is a limited amount of emotional experience, lack of differentiation and adequacy of emotions, indifference and emotional coldness. The amygdala ensures the assessment regulation of behavior depending on the differentiation of emotions. In case of injuries, the ability to assess the emotional side of the mimic is lost, which hinders communication and social interactions.

The amygdala is the structure that organizes vegetative and involuntary motor reactions in fearful behavior. The amygdala nuclei are connected bilaterally with the hypothalamus, the hippocampus, the neocortex and the thalamus. The amygdala receives sensory information from the thalamic nuclei and from the cortex. For some inborn emotional reactions, such as fear, the activation of the amygdala is carried out through the thalamus, that is, faster than realizing the perceptions through the cortex. This prepares the amygdala for the cognitive representation of emotions through the information coming from the prefrontal cortex. The feedback with the cortex leads to the creation of a conscious emotional experience. The rapid activation of the thalamus is biologically appropriate for the organization of behavior under threatening circumstances.

Besides fearsome or negative emotional reactions, the amygdala also mediates the enjoyable emotional responses resulting from the taste of food and satiation of appetite. Together with this accompanying positive emotion, food can serve as a rewarding phenomenon. The amygdala has a key role in creating conditional relationships with an environment in which there is a reward. The amygdala also mediates adaptation to an environment by increasing contact with biologically safe conditions and avoiding harmful ones; direction towards conditions with positive support (food, water, sex) and avoidance of negative environments.

The amygdala also plays a role in learning processes, i.e. in forming memory traces in the amplifying action of negative or positive emotions. Amygdala lesions block the memorization of emotionally significant sensory stimuli. It is associated both with pleasurable and threatening stimuli that activate it and induce stereotyped and relatively simple vegetative and endocrine responses, which prepares the body for some adaptive behaviors.

Amygdala activity is associated with the assessment of the subjective significance of events (Bechara, A., Damasio, H., Damasio, A. R., 2003, Fuster, J., 2015) and emotional memory (LeDoux, J. E., Pine, D. S., 2016, Moore, M., Culpepper, S., Phan, K. L., Strauman, T. J., Dolcos, F., Dolcos, S., 2018). The amygdala is one of the important components of a functional brain system that ensures the processing of socially significant information and social interaction (Frith, C. D., Frith, U., 2012, Carr, J. H., 2011). According to Dolcos (2012), the simultaneous activation of the amygdala and the cortex of the medial part of the temporal lobe, when receiving information, enhances the possibilities for its storage in both short- and long-term memory.

Anxiety and restlessness soothe the limbic system by activating the intermediate prefrontal cortex and reducing amygdala activity. When positive memories are activated, the density of the neurons in the ventromedial and lateral prefrontal cortex changes. Conscious naming of emotions reduces their impact by decreasing the activity of the amygdala and increasing the activity of the frontal cortex. Decision-making reduces the activity of the striatum. According to Korb, A. (2015), hugs and positive tactile stimulation lead to the release of oxytocin, which reduces the activity of the amygdala.

It has been proven that a certain group of neurons in the amygdala are programmed to memorize scary experiences and to alert the body when similar danger signals emerge. The amygdala suppresses all other brain activity during danger. There is the so-called back-up fear center, which is activated when the activity of the amygdala is blocked. It is known that fear distorts visual, auditory and spatial perceptions. Time perception slows down in stressful situations, which is demonstrated as a memory distortion since the amygdala keeps the memory of previous such experiences and overlays the memories, therefore the events are perceived as if they last longer.

The classic experiments (Hess, W. R., Akert, K., 1955) of non-anesthetized animals indicate that characteristic manifestations of emotional reactions, such as fury, can be induced by irritation of hypothalamic areas. The hypothalamus plays a role in emotional reactions since it influences hormonal secretion, the activity of the parasympathetic nuclei in the brainstem and the sympathetic centers in the spinal cord. The pituitary hormones oxytocin and vasopressin are synthesized in the neurons of the hypothalamus. Physiological changes in emotions are due to the hypothalamic control of hormonal secretion.

Neuroscience and Child Brain Development

Developmental neuroscience studies the formation of functional brain organization in ontogenesis, the relationship between child behavior and the formation of brain structures and the relevant mental functions. Developmental neuroscience arises at the border of neuropsychology, psychophysiology, child neurology, and age psychology. From the 1970s of the XX c. till present days enough data have been obtained to corroborate the hypothesis that the child's brain does not function as the adult's brain. The rapid development of pediatric neuroscience began in the 1970s of the XX c. (Hartlage, L. C., & D'Amato, R. C., 2008). At the beginning of its development, it borrowed most of its models from adult neuroscience. Yet, even initial studies clearly demonstrated that these models could not be applied to children.

The formation of individual brain structures and mental functions is a long process that begins as early as the intrauterine period. During pregnancy, between the third and the thirtieth week, the main brain structures are formed. First, the posterior brain regions are formed – the medulla oblongata and the cerebellum. Between the 12th and the 16th week, the mass of the cortex grows and the corpus callosum is formed. In the 17th-18th week, the individual brain centers are formed and they begin to regulate activities. During the 25th week, the six layers of the cortex are distinguishable, which suggests cell differentiation. By the 30th week, myelination begins. After birth, the baby's brain has the potential to develop and ensure individual mental functions; this development, however, may be subject to various risks that prevent its regulatory function. The development pattern of mental processes is heterogeneous and asynchronous, but over a certain period of time asynchronous processes are synchronized in mental activity, which implies an adequate response to the requirements of the environment and society. At the end of their first year, children develop general and fine motor skills that allow them to move around and get to know their surroundings, as well as to pronounce their first words. During this period, the right and left frontal lobes begin to control different spheres of their life (Shaffer, Kipp, 2009). Studies have shown that in girls the left hemisphere is formed first, and in boys - the right (Giedd, 1996). This may explain why injuries in the right hemisphere have far more dangerous consequences for boys while in the left one - for girls (Van der Plas, 2010).

Kolb, B. and Whishaw, I. Q. (2009) have emphasized that the main problem of neuroscience is to determine what is localized in the cortex, as one way to do this is to see how the child's brain is organized and how it works. The smaller the child, the more pronounced is the dynamics of development of the child's brain. Ontogenesis of the brain organization of mental processes goes from the stem and subcortical structures to the cortex, from the right hemisphere to the left, from the posterior brain regions to the anterior ones. Maturity is associated with the downward influence of anterior regions of the left hemispheres on the subcortical ones. Morphological maturity of individual fields and areas is acquired at different times, while the anatomical maturation of the child's brain is a basis for functional ensembles of brain structures that provide mental development.

According to Johnson, M. H., Karmiloff and Smith, A. (2004), neuronal development is under genetic control, but the experience can control the speed of this development. The process of the brain's morphological and functional maturation is a sufficiently long one that increases the risk of negative environmental influences at each stage of its development. It is believed that there are zones in the brain with congenital specialization, others - with variable specialization and third ones that have no specialization (Jonson, 2001, Jonson, Karmiloff-Smith, 2004). Many theories and scientific ideas attempt to explain the changes in cortical and neuropsychological development of children (Jonson, Karmiloff-Smith, 2004). The possibilities for this have increased significantly after the introduction of neuroimaging technologies in clinical practice. The main difference between the child's brain and the adult's brain is the plasticity of the brain connections and of the nervous system. In ontogenesis, the child's brain matures heterogeneously, i.e. unevenly. Different brain structures and the corresponding mental functions mature at a different speed and reach maturity at different stages of child development. This is a regularity of normal ontogenesis; thanks to it each new stage is the result of complex inter-functional reorganizations. Individual variants are formed on this basis and they are manifested in the uneven maturation of structures and functions. The unevenness in higher cortical functions development is a normal phenomenon having an adaptive significance; it is determined by the fact that these functions are of social origin, they are indirect in structure and volitional in the way of functioning. This means that both biological and social factors will play a significant role in the maturation of these functions. The more significant the biological factor is, the more insignificant social factor will provoke the occurrence of a disorder, and vice versa - a significant social factor can trigger an insignificant biological factor. Among the social factors, a significant role is played by the family system, attitudes towards the child, active stimulation of development, etc. The very essence of a child's neuropsychology covers thoroughly the understanding of the bio-mental-social model in the understanding of disorders.

Intrasystemic heterochronicity is associated with the permanent complication of functional systems. Structures that ensure more elementary levels of the system are initially formed, and then new elements appear to increase its efficiency. Each higher cortical function develops both horizontally and vertically. Vertical development is related to the process of automation. Intrasystemic reorganizations mean a change in the hierarchical interaction between the individual units of a system, which leads to qualitative transformations of mental functioning.

Intersystemic heterochronicity is associated with the non-simultaneous formation of individual functional systems and with the changes in the interaction between mental functions during which one or another mental function has a leading role in mental development and functioning. Anatomical and/or physiological pathology in individual brain structures does not always lead to functional impairment in childhood, which is associated with the functional immaturity of the child's brain. Plasticity arises precisely because of the heterochronicity of the maturation of the individual functional systems. As a result, the incomplete process of formation in a young functional system allows for compensation of an already existing defect in the functional system of a previous stage.

The fundamental postulate of child neuropsychology is that the maturation of individual brain structures is the basis for the formation of corresponding mental functions, as the update of the child's brain resources is carried out only in the conditions of external stimulation. This implies the provision of a stimulating environment for the child, especially during the first year of their life, when the attachment to the main care-giver is built, which determines their further social functioning, creation of lasting and satisfying relationships, formation of attitudes towards themselves as an individual who deserves attention and confidence that their needs are valid and others can support their realization.

The basic idea of Bowlby, J. (2008) is that attachment behavior has a biological basis that can only be understood in an evolutionary context. He believes that we have a huge number of behavioral patterns that are flexible in terms of the environment and allow us to cope with a huge number of changes in it. We have relatively few stable patterns of actions, but great plasticity in the learning process during the long period of immature helplessness. Bowlby suggests that once the human species has survived a long period of vulnerability, the human youngsters should be equipped with environmentally stable behavioral systems that, through parental care, serve to reduce the risk of immaturity. It is the attachment behavior that seems to be one of the most stable behavioral systems among the species.

Studies have shown that ambivalent attachment is associated with the anaclitic depression characterized by anxiety from the disintegration of interpersonal relationships, fear of abandonment and loneliness. Avoidant attachment is associated with the introjective/self-critical type of depression, characterized by fears of loss of self-esteem and feelings of uselessness, reproach and guilt (Levy, K. N., Blatt, S. J., 1999).

The attachment system protects the child from potential dangers and plays an important role in the regulation of emotional processes. Initially, this regulation is possible only as co-regulation, i.e., it is carried out with the help of the main caregiver, and at a later stage, the child develops the ability to self-regulate their emotions and behavior. The regulation of emotions and behavior is an integral part of mental health, which is built up in the process of parent-child interaction (Matanova, V., 2008). One of the detrimental consequences of the lack of stable attachment is the child's chronic inability to modulate their emotions, behavioral patterns and impulses. The attachment formed plays an important role in mastering self-control. By contrast, children, who have built a secure attachment model, develop well-functioning strategies for self-regulation of emotions and coping with distress in various life difficulties.

Recent studies confirm the assumption that these interactions form and support the ability for resilience, which is seen as a process where, through interactions between the child and the environment, he or she develops the capacity to successfully adapt to unfavorable conditions (Egeland, B., Carlson, E., Sroufe, LA, 1993; Rutter, J., 2002).

Maltreatment and negligence outline a picture of real physical and mental traumatism that places the child in high-risk situations with regard to the development of attachment disorder. Contemporary neuropsychological studies suggest that love, care, and security are responsible for the development of the hippocampus and its regulatory functions, which provide the individual with an opportunity to survive mentally. In this sense, physical and mental traumatism is also defined as a source of changes concerning neurobiological and neuropsychological development. Physical or mental trauma in early childhood affects the development of certain brain structures related to the regulation of emotions and behavior (Nadeau, M. E., Nolin, P., Chartrand, C., 2013). Shore and Tetrick (Shore, L. M., Tetrick, L. E.) argue that "the brain regions responsible for the early attachment process are 'the most plastic in the brain,' which allows their reorganization throughout the whole life" (J. Cassidy, 2001).

In early maternal-child interactions, the mother is the "regulator and container of the affect, and the mother's response to the affect becomes part of the baby's affective experience" (Koren-Karie, N. I. A., Oppenheim, D., Goldsmith, D. F., 2007). Fonagy and Target argue that 'when the mother reflects or mirrors the anxiety of the child, this perception organizes the child's experience and the child 'knows' what it feels.' This process is difficult for parents who are filled with fears and therefore are unable to work out the fears and anxiety of their child and to respond to their needs. Satisfying primarily the child's physiological needs, the parent deprives them of the status of an entity that has their own desires and needs. When the main caregiver fails to recognize the child's emotions or responds coldly to them or postpones them in time, the child gradually stops showing them and 'shrinks' them, thus the child's emotional repertoire does not develop or remains poor and undifferentiated. This inevitably leads to an inability to model their own emotions; it blocks the development of the ability to understand the other because emotions are incomprehensible to them. Mentalization theory theorizes that insecure attachment representations are associated with deficiencies in the child's ability to decode emotional signals, their own ones as well as those of others. Secure representations imply a better understanding of the other's mental state. P. Fonagy talks about the mentalization capacity as an ability of 'reflective functioning.' It is the ability to think about your own inner states and those of others.

Early emotionally charged communications affect the maturation of the right hemisphere, the so-called 'emotional brain.' Reciprocal visual-facial, auditory-prosodic, tactile-gestural non-verbal communications are the biological core of affection between the child and the main care-giver. These emotional connections expand the baby's developing brain regulatory systems. Schore believes that emotionally charged relationships change the developing right hemisphere, which, for the rest of our life, is dominant for non-verbal, holistic, spontaneous processing of emotional information and social interactions (Schore, A. N., 2018). This enables the body to regulate the impact and challenges of the environment, thus it forms emotional resistance and well-being in the later stages of life (Schore, 1994, 2003, 2012, 2014, 2018). The structure and functions of the brain are shaped by social interactions, especially those that involve emotional connections. "Self-organization of the developing brain occurs in the context of a relationship with another self, another brain" (Schore, 1996).

A number of neuroimaging tests have documented the earlier maturation of the right hemisphere in the prenatal and postnatal stages of human development (Schore, A. N., Newton, R. P., 2013). This indicates that the nature of attachment influences the "early programming of hemispheric lateralization" and the dominance of the right hemisphere in the first year of life (Stevenson, I. H., Rebesco, J. M., Miller, L. E., Körding, K. P., 2008).

Hecht, D. (2014) cites a study of lateralization, which shows that the right hemisphere is the main driver of the innate biological need for belonging and social connectedness, thus it regulates the emotions and personal growth. These functions are formed from early attachment experiences.

Taking into account the hierarchical organization of the brain, Schore suggests that emotional processing in the limbic system is shaped by attachment (Schore, 2000, 2005). The right hemisphere (more than the left one) is connected with limbic and subcortical areas, therefore the functional maturation of the limbic system is significantly influenced by the early socio-emotional experience. The neuronal anatomy of the emotional limbic system is a set of hierarchically organized 'control systems' in the right amygdala and the right orbitofrontal cortex (Schore, J. R., Schore, A. N., 2003, 2008, 2012, 2013).

Neurobiological studies have found that the fastest change in the maturation of the brain occurs during the first 3-6 months of life, followed by a slower change of up to 24 months and a relative stability after 24 months (Hermoye, L., Saint-Martin, C., Cosnard, G., Lee, S. K., Kim, J., Nassogne, M. C., Wakana, S., 2006). The function of the amygdala contributes to the building of a secure attachment (Lemche, E., Giampietro, V. P., Surguladze, S. A., Amaro, E. J., Andrew, C. M., Williams, S. C., Simmons, A., 2006). Coping with stress in early childhood increases myelination of the orbitofrontal cortex, which controls the regulation of excitation and encourages emotional sustainability (Katz, D. H., 2014).

This supports the idea that effective early intervention in periods of increased brain plasticity will help expand the right hemisphere not only in childhood but also throughout the rest of life. "Most mental disorders... start much earlier in life than it was thought before" (Insel, T. R., Fenton, W. S.,

2005). Neurobiologically informed programs for early intervention can help optimize emotional well-being and prevent emotional disorders.

The underfed baby has higher levels of cortisol even as an adult, i.e., its response to stress is programmed in the womb or early childhood. Brain systems that manage emotional responses are formed from early events in the baby's life. They are 'programmed' so as to structure the brain to survive the situation, i.e., the brain adapts to the emotional environment. Brain structures are formed from early childhood experiences and this, in turn, determines future social problems. For example, victimized children are those who show insecurity and lability that have formed from experiences with the main care-giver. Maternal depression may develop a hypersensitive response to stress and lower activity in the left brain hemisphere of the child. Personality disorders are also associated with inadequate early emotional care. The lack of a warm relation around the second year is a prerequisite for a later anti-social behavior (Belsky, J. A. Y., Hsieh, K. H., Crnic, K., 1998).

The prefrontal cortex grows at a considerable rate during the first two years of life. This is the area that builds connections in response to social stimuli in early childhood. Brain development is, in fact, a process of creating, strengthening and eliminating connections between neurons. The prefrontal cortex is the brain area that we call the social brain. This area is activated when we have to control our emotions, to guess others' social signals, when we think about feelings and feel sympathy for others. It evolves in response to the baby's social experiences. Neuronal networks result from the baby's current experience - they need a care-giver who in turn transfers his/her experience related to the efficient management of emotions to them. Babies need this constantly, especially during the first two years of their lives.

During the second and third year of the child's life, the number of neuronal networks gradually shrinks based on the principle "if you do not use it – it disappears." This means that the baby will keep neuronal networks connected with certain emotional experiences. If the baby is raised by negative and aggressive care-givers, it will keep neuronal networks related to anger and aggression. When the main care-giver is not available and does not respond, babies show an increased level of cortisol, which may have quite a few negative effects. Negligence, aggression, rude treatment of the baby induce a physiological response. It has been found out that the levels of cortisol are related to later aggressive behavior (Gunnar, M., Quevedo, K., 2007). High levels of cortisol during the first three years of the child's life affect its brain structure. It turns out that those brain areas are affected that develop during the first years of life. Traumatized children have a reduced volume of some brain structures, especially in the prefrontal cortex. Crucial for the development of the prefrontal cortex is the period from the second half of the first year until the third year when the middle prefrontal cortex develops the fastest and is most susceptible to influence. It starts functioning around the 18th month until the second year when the child realizes the self - they can recognize their image in the mirror and have mastered the basic programs of self-awareness and self-control.

Extremely important for the child's development are positive stimuli. Babies learn to notice their feelings when they obtain enough information about them from their care-givers. The social brain has the potential to influence and control the impulsive reactions based in the amygdala. Research shows that self-control at the age of 22 months is largely proportionate to self-control at the age of three or four years (Kochanska, G., Knaack, A., 2003). In fact, self-control is important not only for behavior management but it also turns out that self-control and self-regulation reinforce the ability to learn (Blair, C., Diamond, A., 2008).

Contemporary psychoanalysis is based on randomized controlled studies and research of brain functions, which legitimizes it in the world of science. The strong affectation connected with experiencing immediate danger and fear of losing the attachment figure suppresses the activity of neuronal networks connected with the process of mentalization and activates reactive behavior (Mihova, 2012).

The notion of psychological flexibility has recently been incorporated in the definition of mental health (Ciarrochi, J., Bilich, L., Godsell, C., 2010). Psychological flexibility is defined as the ability to comprehensively experience the present moment, a behavior that corresponds to the values chosen by the individual, even when the present moment includes emotions, thoughts, memories or feelings that are hard to endure (Hayes, S. C., Luoma, J. B., Bond, F. W., Masuda, A., Lillis, J., 2006). Psychological flexibility is central to mental health (Kashdan, T. B., Rottenberg, J., 2010), while psychological rigidity is of crucial significance to mental vulnerability (Kashdan, T. B., Barrios, V., Forsyth, J. P., Steger, M. F., 2006).

There are significant differences in the aims of neuropsychological diagnosis of adults and children and they are determined by the established differences in the functioning of the child's brain and the adult's brain. The aim of neuropsychological diagnosis in adults is to make a topical diagnosis, i.e., to identify the areas responsible for a particular disorder. Due to the immaturity of brain structures and the high dynamics of the processes in childhood, neuropsychological diagnosis in children is functional. The uneven development of individual brain structures and mental functions, as well as the action of various adverse factors, determine this variability. There are no exact age criteria for the construction of individual mental functions. Each function appears at a certain period of time and it subsequently improves its functioning. Delay in the formation of some mental function affects the others, which is a prerequisite for disharmonious development. Active brain development continues in childhood and adolescence, which suggests that brain provisioning of mental activity is also being reorganized. Neuropsychological diagnostics in children aims at assessing the peculiarities in functioning of mental processes - memory, attention, thinking, speech, perception, volitional acts, intellectual functioning, etc. In the process of assessment, the relationships between mental functioning and child's behavior are analyzed. A significant part of the studies in child neuropsychology focus on the development of gnosis, praxis, speech, lateralization of functions, characteristics of attention and their role in solving cognitive tasks.

A significant place in child neuropsychology is occupied by studies related to the lateralization of mental functions. From an evolutionary point of view, this lateralization is likely to increase the capacity to process information by virtue of the fact that the two hemispheres are specialized in different types of tasks. This should not be absolutized since both cerebral hemispheres are activated in certain tasks, but each of them processes information in a different way. This type of specialization helps perform complex tasks in a shorter time and provides highly specialized activities.

The lateralization of functions is demonstrated by the preference for the predominant use of a hand, foot, eye, ear especially in tasks requiring fine manipulations and treatment. Handedness is formed in early childhood and it is not limited to physical activity only but to modality of thinking. Studies of the lateralization of functions prove that counter-lateralization of functions is not always valid. For example, in about 95% of right-handed individuals speech is localized in the left hemisphere, while in only 18.8% of left-handed individuals it is in the right hemisphere.

Discoveries made in the 50s and 60s of the XX c. postulated that the main difference between the two hemispheres was not the level of thinking, but modality and that the left hemisphere had a verbal-analytic strategy, while the right one – a spatial-synthesizing one. This has led to a new understanding of the functioning of the cerebral cortex and subcortex. It is assumed that the left hemisphere is related to analytical thinking, logical sequence of presentation, details and abstract ideas, categorization and deductive thinking, while the right one - to holistic thinking, synthesis and thinking through analogies. In terms of mathematical skills, the left hemisphere is responsible for precise mathematical operations, and both for numerical comparison and evaluation. This relative lateralization gives grounds for certain aspects of behavior to be attributed to a certain brain hemisphere. The left hemisphere is associated with discipline, rules, planning and goal-setting, rationality, while the right one with artistic and musical abilities, spatial orientation, spontaneity, rhythm and prosody. Brain lesions in the right hemisphere affect the ability to identify and differentiate emotional expression. The practical value of lateralization studies is associated with discussions on educational policies and development programs. A major part of school curricula stimulate almost entirely the left hemisphere. For the harmonious development of children, it is necessary to introduce educational techniques for imagination, creativity and visualization that activate the right hemisphere.

The first attempts to systematically study the disorders of higher mental functions in children dates back to the 1960s. Simernitskaya has proved that disorders and recovery of cortical functions in infancy have certain specificities and regularities. At pre-school and primary school age, the incidence of focal lesions is less pronounced and sometimes it is even missing. At an early age even extensive brain injuries do not cause serious behavioral disorganization and may not result in marked symptoms. Unlike adults, in children, the disorders of mental activity are subject to reverse development in relatively short terms. It is particularly characteristic of head injuries in children. In case of organic brain pathology, the compensatory mechanisms are considerably weaker. The atypicality of clinical manifestations in case of local injuries depends predominantly on the location of the injury. In right-handed individuals, cortical and left-hemispherical foci cause more pronounced atypicality and undifferentiation of manifestations, and in case of subcortical and right-hemispherical foci – these occurrences are considerably weaker.

Certain injuries have been successfully localized, which show a delayed effect. In late-maturing areas of the cortex - frontal and parietal, the consequences occur at the age at which in healthy children the corresponding functions begin to actively participate in the regulation of mental processes. The degree of atypicality of the symptoms significantly changes with age, and this regularity is of different nature for the left and the right hemispheres. For the left hemisphere at a lower age, the symptomatology is undifferentiated and atypical, while the symptomatology of lesions in the right hemisphere gets rougher with the younger age of a child. Early cortical injuries are compensated better and to a fuller extent than lesions in subcortical structures. Lesions in subcortical structures in children resemble lesions in cortical structures in adults.

Data from neuropsychological studies in childhood provide scientific evidence about the abilities of a child's brain at different age stages, which could help make more precise the educational and developmental programs and psychotherapeutic interventions. The data from these studies provide explanatory patterns to the behavioral peculiarities in the period of maturation, which significantly enriches the therapeutic programs for children with developmental disorders. There are the so-called 'Windows of Development' in which a gene depends on a certain type of environmental impact determining its expression. These are the periods of major structural changes in the brain - from 15 months until the age of 4 years, pre-school age and early school age (6-10 years), and sexual maturation in puberty (Ornitz, E. M., 1996). There are data about the relationship between childhood trauma and maturation of brain structures when neuromodulation and physiological reactivity change (Pynoos, R. S., Steinberg, A. M., Aronson, L., 1997), which leads to anxiety and increased attention to external irritants. Psychological trauma in childhood can lead to changes in the mid-brain, limbic and stem structures through certain mechanisms triggered by prolonged fear reactions (Perry, B. D., Pollard, R. A., Blakley, T. L., Baker, W. L., Vigilante, D., 1995). Deprivation and rejection in early years may detain the development of the cortex. This leads to a limitation of modulation of reactions of the limbic, stem and mid-brain structures of fear and danger. Early attachment relationships are internalized and encoded as procedural memory. The concept of the 'Windows of Development' is a good basis for developing methods of impact related to health, adaptive functioning and counteraction to pathogenic factors.

The studies of executive functions in childhood have been neglected for many years, unlike the studies of the same functions in adults. This is probably due to the statement of A. R. Luria that the functional maturity of the prefrontal cortex is reached around adolescence (Luria, 1973). A boost in the study of these functions is given by the use of neuroimaging techniques - they allow us to trace which brain regions are activated when performing certain tasks and what differences and similarities there are in the functioning of the child's brain and the adult's brain.

Children are not born with formed executive functions, but they are born with the potential to develop them. If their relations with the main care-giver and adults do not provide them with a stimulating developmental environment, the formation of executive functions is slowed down. The adverse conditions arising from negligence, abuse or violence may jeopardize certain brain structures and impede the development of executive functions.

Barkley, R. A. (2001) notes that executive functions are "critical to organizing and performing complex behaviors over long periods of time."

Executive functions are central processes that are most closely involved with the organization of actions and behaviors, and along with attention and memory are the foundations for studying and learning.

Executive functions are a set of opportunities that are fully achieved in maturity. Studies have shown that it is too risky to extrapolate the pattern of executive functions in adults to childhood, but this strategy is still applied. Effective executive functioning is a fact in adolescence, a time when neuronal systems of the frontal cortex have reached the ultimate degree of myelination, as in adults. There are executive functions that reach their maturity even at an earlier age. They are gradually acquired, which is associated with biological maturation and, accordingly, with the maturation of brain structures. This is a relatively long period of time and because of that, they are at risk from the effects of adverse conditions and, therefore, of inadequate formation. There is accumulating evidence that frontal regions - responsible for executive functions in maturity - do not participate in the same way while these functions are still developing. Perhaps, much more extensive brain structures and neuronal networks are involved in this process. What is particularly important here is that frontal regions are a regulator of the complex forms of mental activity. Frontal regions control thinking, volitional movements and are responsible for a creative approach in problem-solving. They also control emotions and participate in the formation of the child's personality (DuBuc, T., Martindale, M., 2009). When the child grows up, the frontal lobes will help them plan and organize their everyday life, understand the meaning of judgments, and draw conclusions. It is known that the transition from a sensory-motor level to a higher one is associated with the maturation of prefrontal regions, which is also associated with the improvement of working memory (Bell, M. A., Fox, N. A., 1992).

Executive functions start their development in early childhood and this process goes in parallel with the development of the frontal lobes and other brain structures. The child's growth is accompanied by an increase in the requirements concerning their self-dependence and purposefulness - from elementary self-service to solving complex cognitive tasks, which manifests itself as an inter-influencing cooperation. Self-dependence is possible at a certain level of maturity of brain structures; they, however, are influenced by manifestations of independence and purposeful behavior.

Studies of children show that without the skills to inhibit impulses, to focus and withhold information, intention itself cannot be transformed into behavior (Shing, Y. L., Lindenberger, U., Diamond, A., Li, S. C., Davidson, M. C., 2010).

A major part of behavioral patterns in adulthood are laid down in early childhood. The World Health Organization has published a report by the Commission on Social Determinants of Health, which states: "Health and wellbeing issues in adults - such as mental health problems, weight issues, heart diseases, crime, poor literacy - are often linked to childhood experiences. Investment in early childhood is the most powerful investment a country can make, with returns over the life course many times the amount of the original investment." Often the living conditions in early childhood significantly affect the overall development of the individual.

Chapter 8

NEURONAL BASES OF MENTAL DISORDERS AND SPECIFIC PSYCHOTHERAPEUTIC CONCLUSIONS

Vanya Matanova and Zlatomira Kostova

In its long history, psychotherapy has received quite unconvincing scientific support. The development of pharmacotherapy has a scientific basis and it allows for assessing the effect of a given drug. The approach to psychotherapy could be similar. Making a connection between psychotherapy and neurobiology would help provide neurobiological evidence of the efficacy of various forms of psychotherapy. The lack of such evidence is caused by difficulties for direct observation of neuronal changes in the brain. The development of functional technologies for neuro-imaging allows for non-invasive monitoring of changes in the human brain. These new technologies make it possible to observe training- and learning-related changes in brain activation patterns. Changes in brain functions and neuronal networks due to psychotherapy can now be observed.

Nowadays, the classification of mental disorders is built not on objective criteria, but on subjective symptoms, verbalized by the patient themselves. So we believe that the peculiarities of the clinical picture in adults are determined by personality peculiarities, while in children - by the developmental profile. The objective criteria for the classification of disorders need to be associated with brain functioning, biochemical and morphological changes, etc. New methods of neuroimaging that allow for a detailed study of brain structure and functions make it possible to examine the mechanisms for the occurrence of different disorders. There is an increasing perception that the effectiveness of psychotherapy depends on the individual features of brain functioning. We are witnessing a process in which psychiatry, clinical psychology and psychotherapy are transformed from speculative sciences into neurosciences.

In the field of psychology, more and more studies are related to brain functioning; even in areas such as social psychology, which is traditionally opposed to the brain-behavior relationship. Recent achievements in the field of clinical neuropsychology are various and quite impressive. Together with the social and psychological aspects, they contribute to the enrichment of psychological diagnostics and psychotherapy. More and more mental disorders are explained by neuropsychological and neurological dysfunctions – depression is caused by the dysfunction of neurotransmitters; restlessness and anxiety are caused by a hyper-reactive sympathetic nervous system and dysfunction of the amygdala; excess weight – by dysfunction of the hypothalamus, sleep disorders are associated with damages of the reticular activating systems, autism - by disorders of interneuronal connections, determined by gene mutations, etc. In some mental disorders, there is disruption in the activity of different brain regions or the connections between them, in the absence of obvious injury.

Internal communication of nerve cells allows us not only to better understand what happens but also to create a set of adequate strategies. As Kandel says, to remember something, you need to repeat it, otherwise, no 'spikes' that make the connection between the nerve cells appear. The ability to learn from experience, Kandel writes, undoubtedly is the most prominent feature of human behavior. A person identifies with what they have learned. A lot of the emotional and mental problems are mastered, i.e., they are the result of accumulated experience. Kandel formulated a very important question: How do the biological processes in the brain create mental events and how do social factors model the biological structure of the brain? According to McCutcheon (2006), the environmental impact and the experience of learning in a social environment can also change gene expression.

Still, psychotherapy is based on the dualistic understanding of mind and brain. Reductive physicalism was developed in the 20th century on the basis of several influential trends connected to medicine and psychiatry: eliminative materialism and identity of consciousness. These tendencies impose neurobiological patterns and associate psychological and psychopathological phenomena with lesions and brain processes.

According to Andreasen (1997), the term 'psyche' means the activity of the cerebrum. But the subjectivity of human behavior should not be reduced to chemistry and physiology. Psychic phenomena occur in the cerebrum, but subjective experiences, in turn, affect brain functioning.

During the last few decades, various studies have been conducted on the neurobiological mechanisms in different mental disorders, and especially in depression and the relation of these mechanisms to psychotherapy. Despite the existing controversy, there is an increasing perception that antidepressants and psychotherapy have different neurofunctional mediators and different mechanisms. Most studies focus on the effect of pharmacotherapy and relatively few are those devoted to the effect of psychotherapy. This has led to an unbalanced approach to the treatment of depression, where the focus is not on psychotherapy but on pharmacotherapy; this approach is based on the traditional perception of pharmacotherapy as biological intervention, and of psychotherapy - as psycho-social intervention. Studies on depression confirm that psychotherapy has neural correlates and that psychotherapy is a kind of a biological approach.

Through the use of various experimental approaches, changes in the brain related to experience were found at the cellular and molecular level in animals. The appearance of functional neuroimages, including Single-photon emission computed tomography (SPECT), Positron emission tomography (PET) and functional magnetic resonance imaging (MRI), made it possible to study changes in the brain system by measuring cerebral blood flow or metabolism. Various experimental methods were used to detect the pathophysiological mechanism associated with the neurofunctional regime and dysfunction of the nervous network in depression.

During the 1990s, studies of patients with depressive, anxiety and personality disorders, who showed evidence for changes in the brain as a response to cognitive-behavioral psychotherapy and pharmacotherapy, were conducted. The results showed similar brain changes in both groups of patients, especially in the caudate nucleus. A large part of the studies reported similar changes in the cerebrum both in psychotherapy and in medication treatment. In the research of Goldapple et al., (2004), the response to cognitive-behavioral psychotherapy (CBT) of patients with depression was associated with an increase in metabolism in the hippocampus and the dorsal part of the cingulate gyrus, and a decrease in metabolism in the dorsal, ventral and medial regions of the frontal cortex. This picture differs from the picture of the changes triggered by Paroxetine, which increases metabolism in prefrontal regions and decreases it in the hippocampus and the cingulate gyrus. In a study conducted by Carlson, N. R. (2012), clear differences were noted between the effects of short-term psychodynamic psychotherapy and the use of Fluoxetine in patients with depression.

In his study, Goldapple found out that the mechanisms were different in patients subjected only to cognitive-behavioral therapy in comparison to patients treated exclusively with paroxetine. The patients from the first group showed elevated activities in their hippocampus and the dorsal cingulate gyrus. Goldapple talks about a specific pattern, depending on the therapeutic response to depression. Antidepressants create a bottom-up effect, while cognitive-behavioral psychotherapy (CBT) triggers a top-down effect by reducing cortical processing. These changes help emotional processes and the processing of environmental stimuli and show a therapeutic effect. Ochsner and Gross believe that ventral frontal and limbic hyperactivation causes negative emotions (Ochsner, K. N., Gross, J. J., 2005). There are still a limited number of studies that examine changes in brain activity in different types of psychotherapy (Seminowicz, D. A.,

Shpaner, M., Keaser, M. L., Krauthamer, G. M., Mantegna, J., Dumas, J. A., Naylor, M. R. (2013).

Many clinical, genetic and biochemical studies prove that depression is not a manifestation of just one risk factor or abnormality of one neurotransmitter, instead, it is the result of various genetic, environmental and developmental factors. According to some neuroimaging studies, depression is an outcome of the abnormality of various interactions, rather than of the abnormalities of a single brain area. Depression is defined not as a disease of a single gene, brain region or neurotransmitter system. Rather, the syndrome is conceptualized as a systemic disorder; it is regarded as the net effect of failed network regulation under circumstances of cognitive, emotional, or somatic stress (Carrillo-Roa, T, McGrath, C. L., Zannas, A. S., MQIler-Myhsok, B., Kelley, M. E., Craighead, W. E., Binder, E. B. (2015).

The study of the cortical-limbic system has become important due to its ability to mediate stress response and to play an important role in regulating emotions. Neuroimaging tests in depression showed structural abnormalities of the limbic region, such as atrophy of the amygdala and a decrease in the volume of the hippocampus. Increased activation of the amygdala and reduced amygdala-cingulate gyrus connectivity were observed when subjects were exposed to negative stimulation. The limbic system, including the amygdala, recognizes and classifies emotional stimuli - consciously or unconsciously, and mediates autonomic nervous system responses according to the emotion. There is a reduced activation of the limbic region when emotional stimulation is presented at a conscious level. Neuroimaging tests have been carried out on the frontal cortex, which mediates cognitive-emotional processing. Depressive patients present reduced activation, blood flow and metabolism of the dorsolateral prefrontal cortex and the ventromedial prefrontal cortex during an emotional task.

Research data show that cognitive-behavioral psychotherapy (CBT), Dialectical behavior therapy (DBT), psychodynamic psychotherapy and interpersonal psychotherapy change the functioning of the brain in patients with depression, obsessive-compulsive disorder, panic disorder, social phobia, specific phobia and borderline personality disorder. The hallmark of borderline personality disorder is the excessive affective excitability. Dialectical behavior therapy (DBT) leads to a reduction of the activity towards emotional stimuli in the brain regions that serve them. A study of Schnell and Herpertz (2011) showed that after DBT there was a reduction of the hemodynamics of negative stimuli in the anterior part of the cingulate gyrus, in the temporal and anterior cingulate cortex and in the left insula. According to Mervaala (2000), in the psychotherapy of a patient with borderline personality disorder, the reuptake of serotonin in the medial frontal lobe and the thalamus was lowered, and after a year the results were the same as in a healthy individual.

One hypothesis about the influence of Cognitive-Behavioral Psychotherapy (CBP) on the brain refers to the more effective, top-down control of the limbic structures of the prefrontal regulation system. There is evidence that cognitive-behavioral therapy (CBT) also influences the concentration of thyroid hormones in patients with a major depressive episode (Barlow, 2000, MacQueen, G. M, 2003). The concentration of thyroxine was noticeably lowered and changes in sleep patterns - the same as antidepressants cause - were observed (Thase, M. E, 1997). In out-patients with mild depression, there were no differences in the results between psychotherapy, pharmacotherapy and placebo, although in those with moderate to major depression both approaches exceeded the effect of placebo. Functional magnetic resonance imaging (fMRI) is used to measure brain activity during the period of therapy of depressive patients and the subsequent period, especially to establish brain activity in cognitive tasks.

Contemporary technologies make it possible to confirm that the mental activity of extroverts is higher, which, in turn, is associated with different levels of dopamine. In positive events, increased activity in the amygdala and nucleus accumbens is observed. The biggest difference between extroverts and introverts lies in the processing of different stimuli coming into the brain. In extroverts, the excitation moves in regions responsible for the processing of sensory information, while in introverts, it passes through regions responsible for the processes of remembering, planning and decision-making. These differences necessitate the need to take into account the introversion-extroversion dimension in planning psycho-therapeutic interventions.

It is common knowledge that the unconscious is an integral part of psychotherapy. Recently, neuroimaging has made it possible to understand the unconscious process as well. Studies of depressive patients suggest that they are unaware of their anxiety. This confirms that unconscious anxiety cannot be consciously controlled in patients with depressive disorder. The amygdala also responds to social stimuli that are not consciously registered. The subjects experienced anxiety but were unable to determine its source. This is probably the neurobiological confirmation of Freud's theory that the brain actively responds to stimuli of whose existence it is not even aware.

A study of people with a family history of depression showed that the volume of the hippocampus and the prefrontal cortex was significantly reduced in this group. This result explained some of the hereditary aspects of depression, thus outlining the risk factors for depression. There was a significant increase in the activity of the amygdala, in contrast to the accompanying levels of activity of the putamen and insula, while the activation of the anterior cortex was significantly higher.

Lehto et al. (2003) examined 19 out-patients with depression, 8 of whom with atypical depression, treated by means of psychodynamic psychotherapy for a period of 12 months. Single-photon emission computed tomography was used before and after the psychotherapeutic treatment. The density of serotonin in the mid-brain was significantly increased during psychotherapy in patients with atypical depression, but not in those with classical depression. No changes of dopamine were noted in the striatum. It was found out that recurrences were less frequent in patients who had received psychotherapy than in those treated with antidepressants. Medication treatment had a 'psychological' effect, complementing the immediate impact on the brain's biochemistry, while psychotherapy acted on brain tissue, complementing the drug effect.

There are no data examining the effect of psychotherapy on the symptoms of anhedonia in depression. Such data would provide an opportunity to assess how psychotherapy could increase the functioning of those brain structures mediating the reactions to pleasure.

Neuroimaging also makes it possible to discuss the prognosis of depressive disorder. For example, the prognosis is better in depressive patients with hypermetabolism than in those with hypometabolism of the cingulate cortex. Depressive patients with hypometabolism in the orbital cortex respond better to pharmacotherapy, while those with hypermetabolism in the orbital cortex respond better to psychotherapy.

With the help of neuroimaging, it was found out that a central role in the onset of depression plays a small part in the prefrontal cortex - Brodmann area 25 (BA25). This area is related to the amygdala, which is responsible for fear and anxiety and to the hypothalamus, which is associated with the reactions to stress. These areas exchange information with the hippocampus, which is responsible for memory and with the area of the insula involved in the formation of perceptions and emotions. Thus, Helen Mayberg (2003) proved that in depression the activity of Brodmann area 25 was increased but as a result of psychotherapy this activity was decreased and symptoms were relieved. This area is very rich in proteins that transport serotonin. The amount of neuromediator serotonin in the nerve endings depends on their activity. Lukas Pezawas (2015), Andreas Meyer-Lindenberg & Tost, H. (2012) et al. studied the brain in people with depression and without depression. It was found out that in patients with depression Brodmann area 25 was smaller in volume and its activity was shared with the amygdala. In depressive patients, there was a disruption of the bonds between area 25 and

- the hypothalamus and the brainstem, which manifested as an appetite, sleep, and general activity disorder;
- the amygdala and the area of the insula, which manifested as decreased mood and anxiety;
- the hippocampus, which led to memory and attention disturbances;
- frontal parts, which disturbed critical attitude and self-esteem.

Area 25 controls the perceiving and modulating activity of an extensive network of lower-level centers responsible for fear, memory, and self-esteem. Disturbances in the functioning of BA 25 lead to a disorder in the coordination of the activity of these centers, errors in the processing of information, inadequate assessment with regard to the world and one's own condition. This implies planning of therapeutic interventions associated with these components.

Many studies have been conducted in order to investigate the importance of the amygdala in relation to depressive disorder. Cortex-amygdala and amygdala-cortex circuits influence the time to form a new formation, which serves as a guideline for determining the duration of psychotherapy. Psychotherapy proves effective in the control of implicit memory, therefore it is necessary to encourage implicit memory through free association. The amygdala is activated during free associations, which means that patients should play an active role in psychotherapy, they should create emotional and cognitive associations connected with the trauma. The process of psychotherapy in depression is not a passive conditioned reflex, but an active learning process.

It was found that amygdala activation increases while aversive stimuli are expected. The reaction of the amygdala in patients with depression was elevated during emotional tasks, while dorsolateral prefrontal cortex (DLPFC) activation decreased during cognitive tasks. Researchers also found that the functional connectivity between the amygdala and DLPFC decreased in patients with depression. These findings form the basis for the theory that brain functioning in depressive patients shows increased bottom-up emotional reactivity and lowered top-down emotional reactivity in the regulation of emotions. This is the so-called 'cortico-limbic dysregulation model.' This model, proposed by Beck, best explains the mechanism of the neurofunctional regime and the dysfunctional network in depression.

Another study proved that prefrontal activity was decreased as a result of top-down processing after cognitive-behavioral psychotherapy. The patients showed decreased activity in the orbitofrontal cortex and the left medial prefrontal cortex and increased activity in their right occipitotemporal cortex. The authors of the study came to the conclusion that cognitive-behavioral therapy was related to reciprocal modulation of cortical limbic interactions (Kennedy, S. H., Konarski, J. Z., Segal, Z. V., Lau, M. A., Bieling, P. J., McIntyre, R. S., Mayberg, H. S., 2007).

Fu discussed the effect of cognitive-behavioral psychotherapy (CBT) on the manifestations of depression by using fMRI (Fu et al., 2004). Scans of 16 patients with depression were compared to those of healthy control subjects. Before psychotherapy, the depression patients were exposed to a sad facial expression and the reactivity of their right amygdala and hippocampus was higher than that of the healthy subjects. These differences between the groups were no longer found after 16 sessions of cognitive-behavioral therapy. The results showed that psychotherapy did not change only specific regions, but extensive brain areas and, as a whole, it modifies brain dysfunction.

Martin measured the changes in cortical brain functions after interpersonal psychotherapy (IPT) in depression patients. The results of patients who underwent pharmacotherapy and psychotherapy were compared. The patients in both groups showed clinical improvement with increased blood flow in their right basal ganglia. The patients who underwent psychotherapy showed other changes as well - increased right cingulate activity (Martin, S. D., Martin, E., Rai, S. S., Richardson, M. A., & Royall, R., 2001). The results of the study showed that dorsal and ventral prefrontal cortical metabolism decreased after psychotherapy (IPT), while metabolism in the limbic and paralimbic regions increased, which led to a decrease in the symptoms of anxiety, somatization, tension, and fatigue. The improvement of the cognitive functions was positively correlated with the degree of dorsolateral prefrontal cortex metabolism. Some studies reported that the activity of the dorsolateral prefrontal cortex was negatively correlated with the total score after cognitive-behavioral psychotherapy (CBT). This difference was attributed to the reduction of over-thinking and experience and the increase in general cognitive abilities (Brody, A. L., Saxena, S., Stoessel, P., Gillies, L. A., Fairbanks, L. A., Alborzian, S., Ho, M. K., 2001).

The metabolic activity in the amygdala, hippocampus, and prefrontal cortex of depressed patients treated by means of psychodynamic psychotherapy (PDT) became similar to that of healthy people when patients were exposed to attachment-related stimuli. After psychodynamic psychotherapy (PDT) the activation of the cingulate cortex was decreased in patients with depression compared to healthy people. The change in brain functions was analyzed before and after psychodynamic psychotherapy (PDT), and the results of the control and pharmacotherapy groups were compared. Buchheim (2012) suggested that psychodynamic psychotherapy (PDT) was very effective in patients with major depression, whereas cognitive-behavioral psychotherapy (CBT) did not show such results.

Buchheim et al. (2012) examined the effect of long-term psychodynamic intervention. Sixteen patients with depression who underwent 15 months of psychodynamic therapy were scanned twice - before and after the treatment. During the study, the patients constructed narratives based on scenes from fairy tales, neutral and alternative descriptions, and narratives extracted from an attachment interview. The results showed increased activity in the left hippocampus, the amygdala and the medial prefrontal cortex in the patients in comparison to the healthy controls. The improvement of the general symptoms was correlated with the normalization of brain activity.

Other studies were focused on the changes in neurotransmission related to the pathophysiology of depression following short psychodynamic psychotherapy (Hirvonen et al., 2010; Karlsson et al., 2010). The two groups included in the study showed comparable improvement in their symptoms after treatment, but those who underwent psychotherapy showed increased serotonin binding in several cortical regions including the dorsolateral prefrontal cortex and ventrolateral prefrontal cortex. In a subsequent study of the same group (Hirvonen et al., 2010) the effect of short psychodynamic psychotherapy and Fluoxetine was compared in patients with major depression. Although both treatments led to a significant improvement of symptoms, no effects on receptors in the striatum were registered.

Dichter et al. carried out a study (Dichter et al., 2009) to elucidate the neural correlates of Brief Behavioral Activation Treatment for Depression (BATD). Brain activity was measured in 12 subjects with depression and 15 healthy controls. The patients showed increased activity in structures that mediate responses related to selecting, anticipating and receiving an award, etc. Before and after the therapy the patients were scanned while performing a task requiring cognitive control in both sad and neutral contexts. The results showed that, following treatment, there was decreased activity in prefrontal structures in response to stimuli presented in sad contexts.

During cognitive-behavioral therapy (CBT) patients with depression were reminded of bad or sad memories and were asked to re-evaluate them. Several studies examined the correlations between re-interpretations of the negative emotions and brain activity during cognitive-behavioral therapy (CBT). The patients' status improved after they had re-interpreted their negative memories, which was related to increased activity of their dorsolateral and dorsomedial prefrontal cortex (PFC) and to the decrease in the activity of their amygdala and orbitofrontal cortex. These findings lie at the core of the biological model of cognitive-behavioral therapy (CBT). Negative emotions as a result of certain stimuli appear in the limbic and ventral prefrontal structures, which are decreased or blocked during cognitive-behavioral therapy (CBT).

In terms of depression, the principle of psychotherapy does not follow the simple principle of a decrease in bottom-up regulation and an increase in top-down regulation. The change of activity in the prefrontal cortex also differs between the tonic and resting states.

A number of studies aimed at establishing the peculiarities of brain functioning in obsessive-compulsive disorder. Neuroimaging methods have shown anomalous activity in the orbitofrontal cortex, responsible for decision-making and the system of moral values. In obsessive-compulsive disorder, the activity of sections of the frontal cortex and basal ganglia is increased and more synchronous than in healthy controls. The basal ganglia and caudate nucleus are related to the emergence of obsessive-compulsive disorder. A decrease in the activity of the orbitofrontal region and of the caudate nucleus leads to improvement of symptoms. It is known that there is an innate predisposition to high levels of cholesterol and glucose in the blood, but genetic factors do not cause pathology in themselves - specific interactions with the environment and individual experience are needed for pathology to be triggered. According to Baxter et al, (1998), in obsessive-compulsive disorder which had been treated with a drug and cognitive behavioral therapy (CBT), similar changes in the metabolic rate in the right caudate nucleus were observed. The rate of glucose metabolism was lowered equally in treatment with a drug and in the case of cognitive-behavioral therapy (CBT).

Functional magnetic resonance imaging allows us to connect emotions with certain brain regions and to differentiate various types of anxiety depending on brain activity. For example, anxiety about the future manifests itself as increased activity in the frontal cortical regions, anxiety about public appearances - as increased activity in the anterior cingulate cortex. Anxiety in obsessive-compulsive disorder is associated with disorders of the neural circuit between the frontal regions and the basal ganglia. In other words, neuroimaging can, with great accuracy and objectivity, identify the differences between normal and clinical anxiety; it can also monitor mental activity. Lowering anxiety levels reduces the activity of the amygdala and increases the activity in the frontal lobes.

It is known that subjective experiences change the way of thinking. An example of this relationship is the experience of trauma. Psychic trauma changes not only the chemistry of the brain but also brain structure. These brain structures include the amygdala, the hippocampus and the prefrontal cortex. The brain changes every time we learn something new since the synaptic connections between neurons in the brain are strengthened. This is possible due to the fact that the process of learning engages the so-called 'long-term potentiation' in the cerebral cortex (Fox, 2013).

Studies were carried out with patients with traumas suffered in childhood, as well as studies on risk factors for depression associated with the environment. It was found out that the volume of the hippocampus was reduced in patients who suffered from depression due to sexual harassment before the age of 11. In maltreated children, the volume of the dorsomedial prefrontal cortex was also reduced. These results clearly showed the impact that psychic trauma during childhood had in the pathophysiology of depression. The stimuli associated with the trauma continued to cause fear for a long time. Predisposition for post-traumatic stress disorder (PTSD) is associated with impaired functioning of the ventromedial prefrontal cortex which affects the activity of the amygdala - the generator of fear and anxiety. The activation of these structures is accompanied by reactions to fear. The dendrites of the neurons of the amygdala are responsible for motivation, decision-making, separation of significant irritants, which suggests special therapeutic solutions when working with patients with post-traumatic stress disorder (PTSD). Environmental stress factors can play an important role in the initiation of genetic predisposition to a certain disease in adults. Genetic factors are likely to alter human susceptibility to the effects of psychotic life events. The perception of one's own place in the structure of relationships can affect the activity of neuromediators and their impact on the brain.

In depression, obsessive-compulsive disorder and posttraumatic stress disorder (PTSD), the prefrontal cortex is affected, where goals and motivations are analyzed, where decision-making and planning of activity are carried out. In schizophrenia, there is an anomalous activity of the dorsolateral prefrontal cortex, and in attention deficit hyperactivity disorder (ADHD) the development of the prefrontal cortex in the age range from 7 to 12 is delayed. A study was conducted with 56 outpatients, 54 of whom suffered from paranoid schizophrenia and two - from schizoaffective disorder. All of them were right-handed, without neurological diseases or brain traumas. They were divided into two groups – one of patients undergoing antipsychotic treatment and the other cognitive-behavioral therapy (CBT). It was established that when the patients viewed photographs showing diverse emotions, different sections of their brains were activated:

- the inferior frontal gyrus, the thalamus, the amygdala, and the putamen in the left hemisphere in photographs associated with fear;
- the inferior frontal gyrus, the thalamus - left and right, the amygdala - right, the putamen - left and right, and the lower central occipital region in photographs associated with malice.

No significant changes in brain activity were observed in the patients who received antipsychotics.

There are also studies of molecular and cellular changes due to psychodynamic psychotherapy which test Kandel's hypothesis (1998) that psychotherapy can lead to changes in the expression of genes by means of changing the synaptic connections between nerve cells and by inducing morphological changes in neurons. Studies dealing with the impact of mental processes on the immune system show the effect of mental and social factors on physiological functions which indirectly involve the genetic replication of cells (Ochsner, K. N., Ray, R. D., Cooper, J. C., Robertson, E. R., Chopra, S., Gabrieli, J. D., Gross, J. J., 2004). Changes in biological functions mediated by gene expression related to social experience lead to a change in protein synthesis and eventually to physiological changes (Pinaud, R., 2004). Since psychotherapy and counseling are effective and lead to long-term behavioral changes, they probably act through learning and through changes in gene expression. They change the density of synaptic connections and lead to structural changes that modify the anatomical pattern of nerve cells in the brain (Kandel, 1998).

Muenke (2008) suggests that the therapeutic effects of stress-lowering techniques may be attributed to changes in the expression of genes. This is determined by neuroplasticity, which allows the brain to change its structure and functioning in response to physical and mental activity. When the brain works, it changes. The preservation of cognitive and intellectual abilities in elderly people involved in intellectual work is regarded as a confirmation of this. Mental activity not only arises from the brain, but it also forms it.

Although neurons cannot regenerate themselves as some other organs, there is new evidence suggesting that the brain can engage in neurogenesis, i.e., it forms new neurons (van Praag, H., Zhao, X., Gage, F. H., 2004). These new neurons originate deep in the brain and then they migrate to other brain regions where they form new connections with other neurons (Gould, 2007). This leaves open the possibility that someday scientists might be able to 'rebuild' a damaged brain.

Many studies have demonstrated that the two brain hemispheres specialize in different abilities. Mathematics and assessment of time and rhythm, as well as coordination of complex movements, etc. are localized in the left hemisphere. The right hemisphere, on the other hand, has very limited verbal abilities, and yet it excels in perceptual skills. It is able to recognize objects, faces, patterns, melodies, it can put a puzzle together or draw a picture (Gazzaniga, M. S., 2005). Although Gazzaniga's research has demonstrated that the brain is in fact lateralized and the two hemispheres specialize in different activities, this does not mean that only one hemisphere is used for different activities. Usually, both hemispheres are used at the same time, and the difference between the abilities of the two hemispheres is not absolute (Deouell, L. Y., Sacher, Y., Soroker, N., 2005). The right hemisphere is responsible for non-verbal communication, face-recognition, establishing relationships with others. It processes non-verbal signals that start as early as the beginning of mother-baby interaction. It processes prosodics and the tone of speech we use to transmit our emotions. The development of the right hemisphere is predominant from birth to the age of two years and coincides with the crucial periods in the development of these functions. Creativity requires activation of the right hemisphere, which increases at a time of physical and mental fatigue and implies the generation of new ideas and lateral thinking.

The left hemisphere processes the verbal elements of speech and relies on a conscious analysis of information. By the end of the second year, the right hemisphere is bigger than the left one, which is yet to begin its formation. So, by the age of three years, children are predominantly right-hemispheric. Data from brain scans show that during the first two years of their lives babies and their mothers communicate, above all, non-verbally. The period from 10-12 months to 16-18 months is crucial since it is the time when the right frontal region of the brain develops and neural circuits are formed, which allow the baby to feel a sense of attachment and control of emotions. During the crucial period of emotional development, the mother uses a change of tone and predominantly non-verbal means of communication.

Behavioral Therapy (BT) is related to the desensitization of stimuli that cause anxiety and lead to extinction of learned responses. The results of the studies showed that the regions of the prefrontal cortex (PFC) and the amygdala are related to desensitization responses. This shows that effective behavioral therapies can have biological effects, not just psychological ones (Hollander, E., 1997). These observations confirm the fact that social behavior can be inherited and transmitted through the expression of genes in neuroplastic changes in the brain structure that lead to learning and change. The idea that social experience can lead to changes in gene expression is a key component in the new paradigm of psychiatry, psychology and psychotherapy. This implies the introduction of various techniques for the activation or the suppression of different brain regions, so as to optimize mental activity.

Contemporary studies dealing with the impact of psychotherapy on brain mechanisms take into account chiefly the type of psychotherapy without considering other factors such as the experience of psychotherapist, patient's age, duration of impact, social factors, etc. It has been already proven that there are no strictly specialized cortical regions; also, there is evidence of changes at synaptic, molecular and cortical levels, which provides reasonable grounds for making the psychotherapeutic methods used more precise. Psychotherapy is related to characteristic neural circuitries, and depending on the type of psychotherapy, different neural circuits are activated. It has been confirmed that different types of psychotherapy have different activation processes. For example, in psychodynamic psychotherapy, the therapist gathers information about the patient based on the patient's episodic memory. However, if the therapist uses the free association method, which focuses more on random memories than on episodic memory, the brain regions - frontal, parietal, and temporal - are associated. On the other hand, episodic memory is associated more with Broca's area and the left frontal lobe. Thus, free association is a less censored process and uses an extensive cortex network. This provides scientific evidence for the high efficiency of integrative psychotherapy with a dynamic basis.

Regulation of gene expression in nerve cells by social factors makes all our bodily functions, including our brain functions, subjected to social influence. These social influences are biologically incorporated into changes in the expression of specific genes, in specific nerve cells, in specific brain regions. This socially induced change is transferred through culture (Kandel, 1998).

A great part of psychotherapeutic methods aim at increasing patients' abilities to solve problems related to self-reflection and the regulation of emotional experiences. The brain regions responsible for these functions are the dorsolateral prefrontal cortex, the ventral pre-clear bark, the dorsal prefrontal cortex, the ventral and dorsal sub-regions of the middle prefrontal cortex, the posterior part of the cingulate gyrus, the insulae, the amygdala, and the ventrolateral prefrontal cortex (Frewen and al. 2015).

The empirical data analysis reported brain changes as a result of psychotherapy, which made it possible to construct models that explain the mechanisms behind the changes that result from different psychotherapies. These models can then be compared with the psychological theories of these psychotherapies. There is empirical evidence for the improvement of brain functions through psychotherapy, but the neurobiological foundations of psychotherapy have not yet been proven. Psychotherapeutic treatment of patients with depression results in a normalization of the activation pattern in frontal-limbic circuitry. Psychotherapy may exert its effects top-down by targeting mainly frontal cortical regions and reducing dysfunctional thoughts. The mechanism of psychotherapy for depression is however still limited. The neural mechanism of psychotherapy is widely recognized, but its process cannot be described yet. Psychotherapy without neuroscience is only empirical, and the possibility of its development is quite limited.

The results of psychotherapy and the mechanisms of change associated with its effects are traditionally examined on a psychological and social level by measuring the changes in symptoms, mental characteristics, personality and social functioning. The dichotomized understanding that psychotherapy is a treatment for 'psychologically-based' disorders, while pharmacotherapy is for 'biologically based' disorders is still alive. Over the last several decades, it has become clear that all mental processes derive from mechanisms of the brain. This means that any change in our mental processes is reflected in the functions or structures of the brain. This dichotomy of understanding is unfounded since there is sufficient definitive evidence that subjective experience affects the brain. Brain visualization can help psychotherapists predict which patients will respond to specific ways of impact based on individual patterns of brain functioning. In other words, a certain pattern of brain functioning can predict what the response to a specific psychotherapy will be. Methods of neuroimaging may help to better understand what types and methods of psychotherapy are recommended for individual patients, as well as to specify the psychological theories about the processes of change in psychotherapy.

Chapter 9

BRAIN-BASED TREATMENT - A NEW APPROACH OR A WELL-FORGOTTEN OLD ONE?

Vanya Matanova and Zlatomira Kostova

What is the influence of psychotherapy on brain functioning? Is it possible for us to determine in which type of psychotherapy the most significant changes in brain functioning are observed? If the influence of psychotherapy on the brain is a fact, does this fundamentally change the training in psychology, psychiatry, and psychotherapy and the inclusion of knowledge from the basic sciences?

Psychotherapy can be considered a type of training that stimulates the patient to learn how to change their behavior, thinking, and regulation of emotions. Psychotherapy is much more than an opportunity to talk to someone with good listening skills. Understanding the relationship between psychotherapy and brain functions is stimulating news and changes the traditional thinking about the place of psychotherapy and pharmacotherapy, the relations between them in the process of overcoming mental problems and diseases. Psychotherapy is assumed to be a form of learning which suggests that the uptake of information in the process of psychotherapy leads to a change in the expression of genes, thus changing the strength of synaptic connections. The gene sequence does not change under the influence of the environment, but the ability of genes to direct the synthesis of individual proteins depends on environmental factors and is regulated by their influence (Kandel, 1998). This explains the phenotypic differences between monozygotic twins and discordance of diseases, such as schizophrenia.

Any mental condition is a brain state and any mental disorder is a disorder in the work of the brain. The effect of treatment of mental disorders is associated with an effect on structural and functional changes in the brain (Kandel, 1998). Neuroimaging gives an objective diagnostics of mental disorders, revealing the reasons for their occurrence and therefore allows for the development of more effective methods of treatment and psychotherapy.

With the emergence of Positron Emission Tomography (PET), Single-photon emission computed tomography (SPECT) and Functional Magnetic Resonance Tomography (fMRI), it is possible to trace the work of the brain with minimal risk to the individual. These tools enable us to detect changes in blood flow, electrical activity, biochemistry and brain structure, to understand what happens in the brain during counseling and psychotherapy. Linden found that psychotherapy interventions could change brain functioning in ways that were similar to the changes caused by medicines (Linden, D. E. J., 2006).

The history of psychology, psychotherapy and neuroscience extends from the ancient world to the present day. Several discoveries presuppose a rethinking of the aim and methods of psychotherapy. The authors of these discoveries have received the Nobel prize - Eric Kandel in 2000, John Gurdon and Shinya Amanaka in 2012, and John O'Keefe, May-Britt Moser and Edvard Moser in 2014.

Edvard Moser works in the field of neuroscience. May-Britt Moser and he discovered space neurons. In 2014, they both, together with John O'Keefe, received the Nobel Prize in Medicine 'for their discovery of cells that form a positioning system in the brain,' an internal GPS. In 1971, John O'Keefe discovered the first component of the system. He noticed that a certain type of nerve cells in the hippocampus were always active when the mouse was in a certain place in the room. Other nerve cells were triggered when the mouse went to another place. Thus, O'Keefe came to the conclusion that these 'place cells' created something like a map of the room. It was not until 2005 that the couple May-Britt Moser and Edvard Moser found one more component of the system in the brain. They discovered another type of nerve cell - the so-called 'grid cells' which allow precise positioning and road detection. So, the three of them practically give an explanation of how a person copes in space. It is believed that their scientific works will help various diseases, such as Alzheimer's, for example.

In 2012 Shinya Yamanaka was awarded the Nobel prize for the discovery that cells can alter the course of the disease process in patients with genetic and other disorders such as amyotrophic lateral sclerosis, Rhett syndrome, muscular atrophy, cardiovascular diseases, Alzheimer's disease, spinocerebellar ataxia, Huntington's chorea, etc. These studies are an essential contribution to the neuropsychobiology of human intellect, cognition and behavior and to the diseases of the nerves and the soul, accordingly.

These and many other discoveries about the brain mechanisms of mental processes and plasticity of the nervous system pose new challenges to psychotherapy.

Studies on the effects that psychotherapy has on brain function give us a reason to believe that it can radically change brain functioning in specific brain regions that are related to cognitive control, self-reflection, decision-making based on remuneration, definition of emotional significance of external events, etc. Does it mean that patients will seek an adequate type of psychotherapy rather than a psychotherapist?

In 2004, Goldapple et al. offered a model in which therapy was directed to different regions with differentiated effects - top-down and bottom-up, which led to a change in the prefrontal hippocampal chains. In other words, psychotherapy can initially affect the so-called higher-order cortical functions, i.e., processes associated with the prefrontal cortex, such as working memory, decision-making and executive functions, due to the fact that the initial effects of psychotherapy are cognitively mediated. Downstream changes in cognitive functioning may affect limbic/affective structures, possibly through reciprocal relationships and/or due to the effect of initial remission of symptoms and the accompanying new behavioral models. The activity of neurons is not a reaction, but a means for changing the relationship between the individual and the environment. All new neural connections are formed only on the basis of past experience, and if there is no such experience, the neurons do not form a new grid. Studies allow the building of a model that explains how changes occur in different types of psychotherapy. These models are compared to the theories of different psychotherapeutic paradigms. For example, the mechanism responsible for the effectiveness of cognitive therapy in patients with a large depressive episode can be associated with an increase in the activity of the prefrontal cortex, which is involved in the mechanism of conscious control, while antidepressants act immediately on the amygdala, which is involved in the generation of negative emotions.

In *Psychopathology of Everyday Life* Sigmund Freud said, "I do not believe that any event in whose occurrence my mental life plays no part can teach me any hidden thing about the future shape of reality; but I believe that an unintentional manifestation of my own mental activity does, on the other hand, disclose something hidden, though again it is something that belongs only to my mental life. I believe in external (real) chance, it is true, but not in internal (psychical) accidental events." He spoke about the ability of the psyche to self-analyze as a prerequisite for harmonization of the given condition. If translated into modern language, can it be argued that Freud talked about harmonization of neural connections? A visit to a psychotherapist leads to the formation of new neural networks. In case of change in the point of view, anxiety rises and activates the error detector which stimulates the patient to build a new neural network. If the rise in anxiety does not lead to external changes, does it mean that internal changes begin? When common neural connections have lost their stability, automation has collapsed, the brain begins to form another behavioral pattern that must be accounted for by the psychotherapist. Will this force a change of the conditions and form an idea about the outcome of psychotherapy?

According to K. Anokhin, our memory retrieves information that is not identical to the real event or recollection of an event, and that each time it reconstructs anew what has been recorded once. When reconstructing the past, the previous information about the event is being deleted and the new reminiscence is being superimposed upon it. If something interrupts this process, then the old one is deleted and the new one cannot be superimposed. When a patient is placed in a situation to recall one and the same event or experience over and over again, the trauma of this situation is lowered. Any situation that triggers unusual behavior, causes alarm and this situation continues until an automatic response to the changed conditions is worked out. Our psyche and body are always either in 'survival mode' or 'thrive mode.' When we have anxious, depressive or negative thinking, the survival mode is turned on. Spending too much time in survival mode affects our bodies, and it limits our ability to learn and thrive.

The impact of the environment on brain activity directs the development of dendrites in such a way that they correspond to the cognitive patterns in the formation of mental representations. Neural connections, the limbic system and the autonomic nervous system form contours which are in compliance with the specific experiences of the growing organism. Cells that are excited together, unite together (Schatz, D. G., Oettinger, M. A., Schlissel, M. S., 1992).

Over the past few decades, a number of studies have provided information on how different types of psychotherapy lead to changes in brain functioning. These studies identify which brain regions are affected during and after the therapeutic process and how these changes lead to a decrease in symptoms and general improvement in the brain functioning of patients. For many years, changes in a patient's condition have been achieved with medication. Now we know that subjective experience changes thoughts. Our brain influences our experiences and our experiences affect our brain (Karlsson, H., 2011).

The ability of the brain to change, repair itself and adapt in response to our experiences is related to the so-called neuroplasticity (Fox, 2013, 2014). It was long assumed that our brain stops developing in childhood, but now we know that the brain continues to develop and change throughout our whole life. Our brain creates new neural pathways and connections. The most commonly 'used' neurons develop the strongest and most stable connections. Learning, experiences, memory, trauma or brain damage can lead to biochemical and structural changes in the brain (Fox, 2014). Neuroplasticity refers to the ability of the brain to change its structure and functions in response to experience or damage. Neuroplasticity enables us to learn and remember new things and adapt ourselves to new experiences. A number of studies in neuroscience are connected with neuroplasticity, which is defined as the ability of the brain to change and adapt to significant changes in the body and the environment. This is carried out by means of anatomical and functional reorganization of the nervous system on two levels:

- the cerebral cortex (cortical plasticity) and
- neural networks (neural plasticity).

This reorganization is achieved by means of activating functionally inactive nerve cells, pathways and connections, change in their number, shape and size, formation of new neurons, etc. Brain reorganization is genetically determined and inherent to everyone since birth. Now we can definitely say that mature brain cells can be reprogrammed to become pluripotent, for which in 2012 the Nobel prize was awarded to John Gurdon and Shinya Yamanaka. Neuroplasticity is the ability of the brain to interact with the environment by adding new connections and neurons with time, which allows constant change in response to life circumstances. The volume of the white matter and the connections between brain cells could not make a clear distinction between maturity or immaturity. Studies have shown that although there are clear structural differences between the brain of teenagers and adults (the amount of gray matter is reducing while the amount of white matter is increasing), reaching maturity does not have one and the same timeline of development, in other words, there are differences between individuals in the maturation of different parts of the brain. There is evidence that some brain regions have not reached full maturity even up to 30 years of age.

If we assume life is a trauma or a process of learning something new that can change our brain structure and biochemistry, then what is the share of psychotherapy in this?

Different people respond differently to various treatment and development interventions. For example, people with schizophrenia can benefit from psychotherapy, but pharmacotherapy is almost always necessary. For many people, the combination of psychotherapy and pharmacotherapy is the best approach. We know that cognitive behavioral therapy causes changes that reduce the symptoms of depression and anxiety disorders, but there is significantly less evidence about other types of psychotherapy. How do various types of psychotherapy affect brain functioning? Can other interventions also change brain functioning?

Neuropsychological studies prove that cognition is a key element that links together brain functioning and behavior. It may be argued that meta-representations have an innate cognitive mechanism and this argument supports the connection between biological causes and behavioral symptoms – this connection is carried out precisely by means of this cognitive mechanism. Social and affective disorders are often explained with a lack of empathy. The reason for this is the lack of a concept about emotions, lack of a mechanism to identify and express emotions, which in itself is a cognitive skill!

In order to understand and control their emotions and establish social contacts, children have to repeatedly experience interaction during the critical period and this should be supported in subsequent periods. If the baby is not taught to control his or her emotions, s/he simply shuts them out. This requires psychoeducation of the mother and the main care-giver in order to minimize the adverse influences during the critical period of emotional development and to serve as prevention for the development of psychopathology at a later stage.

The control of some specific bodily functions, such as movement, vision, and hearing, is performed in specific regions of the cortex, and if these regions are damaged, the individual loses the ability to perform the corresponding function. For example, if a baby suffers damage to facial recognition areas in the temporal lobe, it is likely that s/he will never be able to recognize faces (Farah, M. J., Rabinowitz, C., Quinn, G. E., & Liu, G. T., 2000).). On the other hand, the brain is not 'divided up' in an entirely rigid way. Neural networks of the brain have a remarkable ability to reorganize and extend themselves in order to carry out particular functions in response to the needs of the organism and to compensate for damage. As a result, the brain constantly creates new pathways for neuronal communication.

The brain possesses greater plasticity than many other organs, which requires the formation of a neurobiological approach in psychotherapy in order to reflect the dynamic nature of the interaction between genes and the surrounding environment (Gabbard, G. 0., 2000). The level of plasticity of our brains is highest when we are young children but neuroplasticity continues to be observed even in adults (Kolb, B., Fantie, B., 2013). The principles of neuroplasticity help us understand how our brains develop to reflect our experiences. For example, famous musicians have a larger auditory cortex compared with the general population (Bengtsson, S. L., Ehrsson, H. H., Forssberg, H., Ullen, F., 2005).) and require less mental activity to move their fingers over the keys than do novices (MQnte, T. F., AltenmQller, E., Jancke, L., 2002). These observations reflect the changes in the brain that follow our experiences. Plasticity is also observed when there is damage to the brain or to parts of the body that are represented in the motor and sensory cortexes. For example, a tumor in the left hemisphere affects language functioning, but the right hemisphere will start to compensate in order to help recover the ability to speak (Thiel, A., Schumacher, B., Wienhard, K., Gairing, S., Kracht, L. W., Wagner, R., Heiss, W. D., 2006).). If a person loses a finger, the area of the sensory cortex that previously received information from the missing finger will begin to receive input from adjacent fingers (Fox, 1984).

The brain can continually increase its capacity through appropriate training, which is more effective in childhood when its reorganization is physiological and associated with its maturation. Stimuli that threaten the brain, but do not kill it, trigger processes of endogenous neuroprotection - it is when brain tolerance to the re-enaction of the same stimulus is built.

Brain self-organization is highly individual. In cases of insufficient effective brain capacity, compensatory mechanisms occur; they restore the impaired function, but with changed characteristics. Even in cases of severe brain lesions, a new functional brain architectonic is created over time, where the unaffected cerebral hemisphere becomes more involved. Functional brain recovery does not always correlate with the severity of morphological brain damage. Higher cortical functions are most difficult to restore - gnosis, speech, praxis, memory, etc. By means of a specialized impact-based approach aimed at increasing the brain's plasticity and potential of the brain for reorganization, maximum recovery of impaired brain functions can be achieved.

Words, thoughts, and speech can modify human DNA and influence brain neuroplasticity and reorganization by capturing specific sound vibrations. It is assumed that the information is imprinted in the wave genome (wave genetic program) and can change the heredity and the program of each cell. This principle, which has been applied for centuries, is essential in the non-medicated treatment of all diseases and should not be overlooked these days, regardless of the technological progress of modern medicine. By stimulating brain plasticity, we can entice the aging brain to adapt, while a pathologically changed brain has the potential to reorganize as well as to compensate for the functional deficit that has occurred.

Knowledge about the brain is continually evolving, which leads to periodic updating of medical approaches to various diseases of the nervous system. The discovery of the brain's navigation system by John O'Keefe, May-Britt Moser and Edvard Moser has led to the reconsideration of a number of aspects related to the cognitive processing and space navigation in dementia and cognitive disorders. Through special populations of cells, place cells and grid cells, located respectively in the hippocampus and the entorhinal cortex, the human brain builds the so-called 'Internal GPS,' in which place cells are responsible for the dynamically changing specific localizations and grid cells - for the instantaneous navigation of body position, direction, and speed of movement. The clinical use of this discovery, however, is yet to be examined.

The assessment of the patient's profile and their resources is always subjective. The standard of mental and social functioning is always subject to social and cultural factors, but it is also subject to genetic and physiological factors. For now, the relationship between narrative and physiological markers is rather an exception and is not validated in routine psychiatric and psychological clinical practice (Insel, T. R., Cuthbert, B. N., 2010, 2015). The main problem is to formulate a translation from neuroscience to psychopathology and vice versa. According to Stoyanov (2018), a transfer of terminology and data between the spheres of common interest is needed, which includes two aspects:

- Metalinguistic compatibility. It refers to the commensurability of knowledge and methods applied by different disciplines and implies uniformity of terminology at a meaningful level in neuroscience, social sciences, the humanities, clinical psychology and psychotherapy.
- Axiomatic stability. It implies clarification of the basic mechanism that triggers the corresponding phenomenon. What is the initial cause of mental disorders - the patient's subjective experience or neural processes? (St Stoyanov, D., 2018).

The narrative-based psychotherapeutic method, comprised of the patient's subjective information, is a rather reductionistic opinion on the problem and is not subject to validation beyond the evaluation and analysis of the 'narrative.' Current behavior is determined simultaneously by the traces extracted from the memory of different ages and formed at different stages of individual development. Objective study of the psyche is possible only by studying brain mechanisms. Until recently, it was only possible to monitor the external characteristics of behavior. Contemporary psychological terminology corresponds to this approach and has not changed in centuries. It is associated with the statistical representation of perception, attention and memory by means of characteristics of external behavior. To make an inference about the implicated internal mechanisms based on external behavior, it is necessary to know the relationship between this behavior and brain functioning. Thus, it would be possible to understand the subjective world, not just that part of it the patient is ready to and wants to talk about. The other part is unavailable and cannot be told.

The brain is capable of directing itself towards new sensory stimuli and information. This ability has its biological basis. When new information becomes available, the nerve connections stimulate a state of intense response in the limbic system and the hypothalamus, associated with memory and the mechanisms of pleasure and encouragement. Recurrent situations reduce the activity of *locus coeruleus* and lead to relaxation, languidness, and sleep. New and exciting information increases the activity of brain structures, which is often not accounted for in different types of psychotherapy. Meditation and traditional psychotherapeutic sessions cause certain structural changes in brain physiology, as real as in pharmacotherapy.

During the last few decades, several tendencies have been developed focusing on experimental confirmation of the hypothesis that psychotherapy affects nerve substrates. Solms' research has led to an interdisciplinary interaction called 'Neuropsychoanalysis' (Solms, M., Turnbull, O. H., 2011). This provokes an interpretation of the physiology of the mind and brain, known as 'Double-aspect monism' which perceives the human psyche and brain mechanisms as two manifestations of one and the same basic substrate and can be regarded as close to Donald Davidson's 'abnormal monism.' Neuropsychoanalysis claims that it focuses basically on Freud's *Project of Scientific Psychology.*

Whole generations of psychotherapists were trained to observe professional hygiene by following a particular school and paradigm. Given the diversity of psychotherapeutic approaches, nowadays integrative psychotherapy is becoming more and more widely recognized, without allegations of any negative element of eclecticism. Different psychotherapeutic schools offer different sets of tools, but the effectiveness of psychotherapy does not differ significantly regardless of the theoretical basis (Wampold, 2011). The effectiveness of structured valid therapies does not depend significantly on the theoretical focus (ARA, 2013). Studies have shown that psychotherapists often combine several paradigms and do not follow just a single school (Wampold, B. E., Carlson, J., 2011, 2012, 2013).

The psychotherapist's personality is not an example of a standard and good health and is not the only 'medicine' and is not a major factor in the effectiveness of psychotherapy. The illusion of a psychotherapist's omnipotence is significantly undermined. Preliminary and concomitant therapy and supervision of the psychotherapists themselves is rather an exception than routine practice. The vast number of concepts, theories and methods show that there is no such thing as a unified guiding concept, which, in turn, makes the validity of psychotherapy as a scientific discipline rather questionable. The data from numerous studies often indicate one approach or another as valid and evidence-based, which makes the situation even more complicated. Newer and newer methods appear and for each of them, there are always satisfied and dissatisfied patients. A meta-analysis of psychotherapeutic results carried out in 1994 (Grawe, Donati, Bernauer, 1994) showed that psychotherapy was not a significant factor in providing assistance. The results of Grawe et al.'s meta-analysis showed that a key factor for the effectiveness of psychotherapy is the cooperation between the psychotherapist and the client, or the so-called 'Therapeutic alliance.' But then, doesn't the psychotherapist consider their own images, perceptions, memory, concepts and logical connections, that is, do they not work with the model of their own experiences, and not so much with the patient's issues? The aim of each psychotherapeutic supervision and training is to «neutralize» this influence, i.e., the therapist has to learn to work with the transfer and counter-transfer, something that is regarded as their main competence. The transfer is partly related to procedural memory. What is revealed in the relations with the psychotherapist is a stereotyped, automatic, usual way of constructing objective relations determined to a large extent by the attachment relations during the first years of life. These models recorded in procedural memory are implicit as much as they are unconscious. Psychological protections are a form of procedural knowledge that is encoded in the regulation of affective conditions associated with internalized objective relationships.

Quite a few psychotherapists are poorly acquainted with the achievements of the fundamental sciences because their training is focused on the accomplishments of private sciences. Knowledge in the fields of psycholinguistics, cultural studies, rhetoric, psychophysiology and especially in the field of neuroscience are sparse and unsystematic. The extremely vigorous development of neuroscience requires the systematization of its contemporary achievements and their integration into psychotherapy. The clinical, or also called medical, model in understanding mental issues is stable but does not offer an opportunity for effective psychosocial rehabilitation. Psychotherapists who follow this model often help doctors without being able to take responsibility for the independent overcoming of the mental issues of their patients. On the one hand, this is related to the unclear status of psychotherapy and, on the other hand, to the fact that there is still a parallel between somatic and mental diseases, which makes their treatment basically medicamentous. Thanks to the achievements of neuroscience, a new era in psychotherapy is being discovered, which implies the development of new methods of impact on specially selected brain areas.

Eisner (2000) even discusses the 'death of psychotherapy,' saying that contemporary psychotherapy has become an intellectual pastime or a hidden form of 'brainwashing.' It is necessary to create a system of psychotherapy that rests on the achievements of neuroscience - thus, it would be possible to measure its effectiveness and it would be turned into a discipline based on evidence. This implies the integration of the psychotherapeutic impact and the achievements of neuroscience into a unified system. So, here is the relevant question: What is more important for psychotherapy - the psyche or the personality and can we divide the aim of the impact in such a way? The main task of psychotherapy remains the testing of reality and the creation of an adequate image of reality. And in that sense, what happens in the brain is a relevant question that can be answered with the help of neuroimaging techniques.

The term «Psyche» is an abstract notion, etymologically related to the concept of «Soul». Impact on the psyche is actually an impact carried out upon classes of phenomena to non-identifiable units - for example neurons or an impact on individual elements of the psyche - for example, memory, or on personality. If we choose the second approach we will have to lean only on the patient's narratives, what s/he wants, can or is ready to tell us. In this case the psychotherapist analyzes the narrative at a procedural and meaningful level, but with the help of their own notions and beliefs. The principles of brain functioning will allow psychotherapy to become personalized - only for the individual patient, only for their mental issues.

For a relatively long period of time mental functioning was mainly associated with a personal profile, while brain functioning was laid aside. After the 90s of the 20th century or the so-called 'Decade of the Brain,' today, contemporary specialists work on the boundary between fundamental science and medicine. This brings neuroscience, neuropsychology, psychiatry and psychotherapy closer to each other. The most valuable discoveries about the human psyche have been obtained not within the frame of psychology and psychiatry, but thanks to the fusion of these disciplines with the biology of the brain. According to Kandel, the major efforts are focused on the demand for remedies necessary for the treatment of brain function disorders. In his opinion, the psyche is inherent to and inseparable from the brain, any mental function of the brain - from reflexes to creative forms of activity, are performed by special neural networks in different brain regions. That is why the biology of the human psyche is, in fact, the biology of mind.

Back in 1998, Kandel formulated 5 principles which were revolutionary for his time:

- All mental processes derive from the brain, i.e., have a neural basis;
- Genes are important determinants of the pattern of interconnections between neurons and their functioning;
- Experience modifies gene expression;
- Learning changes neural connections;
- Psychotherapy induces changes in gene expression.

Our brain represents an interaction of neurons that form dense connections, with a defined anatomical structure. E. Kandel proved that our neurons form among themselves additional surfaces of synaptic connections and in this way nervous excitation occurs in them.

Freud and Kandel are the two big names that have changed psychotherapy. The two of them had experienced quite different professional development - Freud started from neurology and biology and ended up with psychoanalysis, while Kandel started from psychoanalysis and ended up with neurophysiology and biology. The great Sigmund Freud took on a new direction - from a neurologist he turned into a psychoanalyst. In these changes, there is more continuity than an inexplicable and strange transformation. He obviously used his twenty years of experience of studying the nervous system in his further psychoanalytic practice.

Today we definitely know that brain structures are being built and altered thanks to experience. Psychotherapy can be more effective when based on a neuropsychological approach – this implies the identification of the neural basis of various disorders and will lead to specific psychotherapeutic conclusions.

Many psychotherapists are excited about the question if there will be a threat to psychotherapy if every psychopathology turns out to have a biological foundation? There is increasing recognition that psychopathology itself is a brain dysfunction or damaged structure. Brain-based treatment stems from the understanding that therapy of mind and behavior changes the body and the brain, whereas the treatment of the body and brain changes mind and behavior!

The XXI century marks a new era of knowledge related to the brain and its functioning! But this knowledge has not yet been integrated into psychotherapeutic approaches. The oft-cited bio-psycho-social model is not sufficiently understood and results in separation of approaches - pharmacotherapy and doctors deal with the biological one, psychologists and psychotherapists deal with the mental one and all the supporting professions - with the social one.

What has always given hope to both clinicians and patients are the incredible phenomena of brain functioning and brain plasticity. Neuroplasticity is the ability of the brain to change and adapt to changes that occur in the body and environment by means of anatomic and functional reorganization of the nervous system. Brain reorganization is genetically embedded and is a mechanism for continuous adaptive change. It is now known that adult brain cells can be reprogrammed so that they become pluripotent. There is evidence of the presence of stem cells in the adult brain which demonstrate ability to form new nerve cells of high learning capacity at any age. This gives reason to believe that, in case of an adequate impact, the brain can continually increase its capacity. The ability of the brain to self-organize, however, is too individual and relatively limited. Optimization of impaired mental functions can be achieved by using specialized impact-based approaches aiming at the increase of brain plasticity and its potential to re-organize.

Knowledge about the brain is being continually enriched, which leads to periodic rethinking and updating of the therapeutic approaches to various diseases of the nervous system and brain dysfunctions. The discovery of the brain navigation system has led to reconsideration of a number of aspects related to cognitive process. There is also a significant amount of scientific evidence suggesting that words and thoughts can modify human DNA and affect the brain neuroplasticity by capturing specific sound vibrations. It is assumed that information is imprinted in the wave genome and this can change the heredity and the program of each cell. This statement, which has been already scientifically proven, has been a fundamental principle in psychotherapy for centuries.

Inadequate behavioral patterns create certain neural connections that determine our behavior and functioning. The repetition of the model is a result of an automated and rooted 'message' from the brain, by means of activating certain neuronal networks. Shifting the focus to new attitudes and beliefs leads to different neural connections and thence to a new reality.

The encouragement of "brain changes" is now associated not only with survival but with personal development and improvement of the quality of life. This is an innovative content, a new theoretical model based on solid scientific evidence. For many years neuroscience has provided psychology and psychotherapy with important information. Changes at the neurochemical and the neurostructural levels affect neuronal networks and may lead to onset of psychopathology. When formulating the type of pathology, it is necessary to take into account the subjective events in the individual consciousness – brain structures and lesions, lateralization, bioelectric activity of the brain, psychological flexibility, etc. But we also have to consider the cultural tendencies and social structures that are inseparable from the development of individual consciousness. That is, each pathology is the tip of the iceberg which comprises brain functioning, the stages of the Self, cultural influences, social structures and spiritual access to them. Psychological flexibility encompasses a wide range of individual abilities for recognizing and adapting to different situational requirements, change in attitudes or behavioral repertoire, maintenance of a balance between important vital areas, openness and commitment demonstrated in behavior which corresponds to certain individual values. In various forms of psychopathology, these processes of psychological flexibility are absent or reduced. The knowledge associated with emotional regulation, acceptance, awareness, social and personal psychology and neuropsychology is integrated in order to create a multi-axis understanding (Kashdan, T. B., Rottenberg, J., 2010). According to the relational frame theory (RFT) of Hayes et al. (2006), psychological flexibility is defined as behavior that encompasses awareness, acceptance and existence, in accordance with values. The lack of psychological flexibility is associated with emotional distress, anxiety, depression, unsuccessful contextual control over cognition and language (Hayes et al., 2006).

All forms of psychotherapy - from psychoanalysis to behavioral psychotherapies - are successful to the extent that they are able to improve the relevant neuronal circuits. Recent theoretical achievements in the neuroimaging methods indicate that the brain is an organ built by experience. For a psychotherapist to be effective enough it is important to have a piece of basic knowledge in neuropsychology, neuroscience and neurobiology. Individual brain structures and regions are related to problems, passions and aspirations. The development of contemporary clinical psychology and psychotherapy is the result of new information about the functions of brain structures.

The study of consciousness, emotions, and imagination were once within the framework of philosophy, psychoanalysis and studies in the humanities. Presently, they are already in the limelight of leading neuropsychological and neurobiological laboratories around the world. This has led to an explosion of new insights into the natural laws that govern our mental world. This does not mean that the mind is associated only with neurobiology and neuropsychology, but that we are free to believe in the power of the mind. The information provided by neuroscience soon will allow us to argue with the classics in the field of psychiatry, psychoanalysis and psychology about the relationships between emotions, experience, will and creativity.

The Concept of 'The Unconscious' still occupies a central place in many theoretical frameworks and clinical approaches, and so it will be. Until recently, no particular attention was paid to the neuropsychological aspects of unconscious processes, but now we have already had data about that. The story of why psychotherapy works is a story of understanding the brain mechanisms of psychic processes, a story of how the brain has been evolving to ensure learning, forgetting and the mechanisms of permanent psychological change.

The new data on brain functioning necessitate the integration of the achievements of neuroscience and neuropsychology in the psychotherapeutic process. An integrative approach is needed to take into account the dynamic interaction between brain functioning, mind, soul, spirit and social interaction, i.e., the development of a model of psychotherapeutic work based on cerebral plasticity!

Brain-based treatment uses neurobiological, cognitive, emotional, social and ecological information in psychotherapeutic interventions. This implies a solid knowledge about the neurobiological foundations of mental conditions and behavior. The methods of psychotherapy can be improved by using the findings of neuroscience and neuropsychology. These findings can be understood if interpreted in the context of psychotherapeutic conceptualization. Brain-based treatment is based on the combination of competencies in the field of neuroscience, neuropsychology, and psychotherapy.

Brain-based treatment aims at changing brain functioning not directly but through experiences and cognitions. Brain-based treatment is the actual bio-psycho-social paradigm leading to integrative psychotherapy. This is neuropsychologically informed psychotherapy. It uses knowledge from neuroscience and neuropsychology to treat cognitive and emotional disorders.

As Eric Kandel believes, learning and memory are extremely important for psychoanalysis and psychotherapy. In any case, many psychological issues are connected to learning, and according to him, 'psychoanalysis is based on the fact that what you can learn, you can also unlearn!' Training and memory are of fundamental significance to our personality as a whole. They make us who we are. In his opinion, all experimental events, including psychotherapeutic interventions, affect the structure and function of neural synapses. This is the same Dr. Kandel who was awarded the Nobel Prize in 2000 and who at the beginning of his professional career was tempted by psychoanalysis!

REFERENCES

Adriani, W., Chiarotti, F. & Laviola, G. (1998). Elevated novelty seeking and peculiar damphetamine sensitization in periadolescent mice compared with adult mice. *Behavioral Neuroscience, 112*(5), 1152-1166. http://dx.doi.org/10.1037/0735-7044.112.5.1152.

Ainsworth, M., Blehar, M. C., Waters, E. and Wall, S. (1978) *Patterns of Attachment: A Psychological Study of the Strange Situation.* Hillsdale, NJ: Erlbaum.

Allen, J. & Fonagy, P. (2006). *Handbook of mentalization-based treatment,* Ed. Wiley & Sons Ltd.

Allen, J. & Miga, E. (2010). Attachment in adolescence: A move to the level of emotion regulation. *Journal of Social and Personal Relationships, 27*(2), 181-190. https://doi.org/10.1177/02654075093 60898.

Allen, J. (2003). Mentalizing. *Bulletin of the Menninger Clinic*, *67*(2), 91–112.

Allen, J. P. (2008). The attachment system in adolescence, In Cassidy, J. & Shaver, P. *Handbook of attachment: theory, research and clinical applications*, The Guildford Press.

Allen, J. P., Insabella, G., Porter, M. R., Smith, F. D., Land, D., & Phillips, N. (2006). A Social – Interactional Model of the Development of Depressive Symptoms in Adolescence. *Journal of Consulting and Clinical Psychology*, *74*(1), 55–65. https://doi.org/10.1037/0022-006X.74.1.55.

Amini, F., Lewis, T., Lannon, R., et al (1996) Affect, attachment, memory: contributions towards psychobiologic integration. Psychiatry, 59, 213-239.

Ammaniti, M. (1988). Transformations of the Self in Adolescence. *Rivista di Psicoanalisi*, 34, 798-820.

Amodio, D. & Frith, C. (2006). Meeting of minds: the medial frontal cortex and social cognition. *Nature Reviews, Neuroscience*, *7*, 268-277.

Andreasen, N. C. (1997). Linking mind and brain in the study of mental illnesses: a project for a scientific psychopathology. *Science*, *275*(5306), 1586-1593.

Armsden, G. C., & Greenberg, M. T. (1987) The Inventory of Parent and Peer Attachment: Relationships to well-being in adolescence. *Journal of Youth and Adolescence*, 16 (5), 427-454. https://doi.org/10.1007/BF02202939.

Ary, D. V, Duncan, T. E., Biglan, A., Metzler, C. W., Noell, J. W., & Smolkowski, K. (1999). Development of Adolescent Problem Behavior. *Journal of Abnormal Child Psychology*, *27*(2), 141–150.

Asen, E., & Fonagy, P. (2012). Mentalization-based Therapeutic Interventions for Families. *Journal of Family Therapy*, *24*(2011), 347–370. https://doi.org/10.1111/j.1467-6427.2011.00552.x.

Asey, B. J. C., & Ones, R. E. M. J. (2008). The Adolescent Brain. *Annals of New York Academy of Sciences*, *126*(1124), 111–126. https://doi.org/10.1196/annals.1440.010.

Astington, J. W., Harris, P. L. & Olson, D. R. *Developing Theories of Mind*, pp. 295-314, Cambridge: Cambridge University Press.

Baillargeon, R., Scott, R. & He, Z. (2010). False-belief understanding in infants. *Trends in Cognitive Sciences, 14*(3), 110-118. https://doi.org/10.1016/j.tics.2009.12.006.

Baird, A., Fugelsang J. & Bennett, C. (2005). What are you thinking: an fMRI study of adolescent decision-making. *Poster. MIT Press Five Cambridge.*

Banfield, J. F., Wyland, C. L., Macrae, C. N., Munte, T. F., & Heatherton, T. F. (2004). The cognitive neuroscience of self-regulation. *Handbook of self-regulation: Research, theory, and applications*, *6283*.

Banks, S. J., Eddy, K. T., Angstadt, M., Nathan, P. J., & Phan, K. L. (2007). Amygdala–frontal connectivity during emotion regulation. *Social cognitive and affective neuroscience*, *2*(4), 303-312.

Barkley, R. A. (2001). The executive functions and self-regulation: An evolutionary neuropsychological perspective. *Neuropsychology review*, *11*(1), 1-29.

Barlow, D. H., Gorman, J. M., Shear, M. K., & Woods, S. W. (2000). Cognitive-behavioral therapy, imipramine, or their combination for panic disorder: A randomized controlled trial. *Jama*, *283*(19), 2529-2536.

Bartova, L., Berger, A., & Pezawas, L. (2010). Is there a personalized medicine for mood disorders? *European archives of psychiatry and clinical neuroscience*, *260*(2), 121-126.

Bateman, A., & Fonagy, P. (2004). Mentalization-Based Treatment of BPD. *Journal of Personality Disorders*, *18*(1), 36–51.

Bateman, A., & Fonagy, P. (2008). Mentalization-Based Treatment for BPD Mentalization-Based Treatment for BPD. *Social Work in Mental Health*, *6*(1–2), 187–201. https://doi.org/10.1300/J200v06n01.

Bateman, A., & Fonagy, P. (2010). Mentalization based treatment for borderline personality disorder. *World Psychiatry*, *9*, 11–15.

Bateman, A., & Fonagy, P. (2013). Mentalization-Based Treatment Mentalization-Based Treatment. *Psychoanalytic Inquiry A Topical Journal for Mental Health Professionals*, *33*(6), 595–613. https://doi.org/10.1080/07351690.2013.835170.

Baxter, L. R., Schwartz, J. M., Bergman, K. S., et al (1992) Caudate glucose metabolic rate changes with both drug and behavior therapy for obsessive-compulsive disorder. Archives of General Psychiatry, 49, 618-689.

Baynes, K., & Gazzaniga, M. S. (2005). Lateralization of language: Toward a biologically based model of language. *The Linguistic Review*, *22*(2-4), 303-326.

Bechara, A., Damasio, H., & Damasio, A. R. (2003). Role of the amygdala in decision-making. *Annals of the New York Academy of Sciences*, *985*(1), 356-369.

Beeghly, M. & Cicchetti, D. (1994). Child maltreatment, attachment, and the selfsystem: Emergence of an internal state lexicon in toddlers at high social risk. *Development and Psychopathology, 6*(1), 5-30. https://doi.org/10.1017/S095457940000585X.

Bell, M. A., & Fox, N. A. (1992). The relations between frontal brain electrical activity and cognitive development during infancy. *Child development*, *63*(5), 1142-1163.

Belsky, J. A. Y., Hsieh, K. H., & Crnic, K. (1998). Mothering, fathering, and infant negativity as antecedents of boys' externalizing problems and inhibition at age 3 years: Differential susceptibility to rearing experience? *Development and psychopathology*, *10*(2), 301-319.

Belsky, J., & Hsieh, K. H. (1998). Patterns of marital change during the early childhood years: Parent personality, coparenting, and division-of-labor correlates. *Journal of Family Psychology*, *12*(4), 511.

Bengtsson, S. L., Ehrsson, H. H., Forssberg, H., & Ullén, F. (2005). Effector-independent voluntary timing: behavioural and neuroimaging evidence. *European Journal of Neuroscience*, *22*(12), 3255-3265.

Bergen, D. L., Schroer, J., & Woodin, M. (2017). *Brain Research in Education and the Social Sciences*. Taylor & Francis.

Bevington, D., Fuggle, P., Fonagy, P., Target, M., & Asen, E. (2013). Innovations in Practice: Adolescent Mentalization-Based Integrative Therapy (AMBIT) – a new integrated approach to working with the most hard to reach adolescents with severe complex mental health needs. *Child and Adolescent Mental Health*, *18*(1), 46–51. https://doi.org/10.1111/j.1475-3588.2012.00666.x.

Bijsterbosch, J. D., Ansari, T. L., Smith, S. M., Gauld, O., Zika, O., Boessenkool, S. L., Bishop, S. J. (2018). Stratifying patients with affective disorders by resting limbic connectivity predicts cognitive bias. *NeuroImage: Clinical, 19*, 425–433. Doi: 10.1016/j.nicl.2 018.0 4.033.

Bion, W. (1962). *Learning from Experience*. London: Heinemann.

Blair, C., & Diamond, A. (2008). Biological processes in prevention and intervention: The promotion of self-regulation as a means of preventing school failure. *Development and psychopathology*, *20*(3), 899-911.

Blakemore, S. (2008). The social brain in adolescence. *Nature Reviews, Neuroscience, 9,* 267-277.

Blakemore, S., & Choudhury, S. (2006). Development of the adolescent brain: implications for executive function and social cognition. *Journal of Child Psychology and Psychiatry*, *47*(3/4), 296–312. https://doi.org/10.1111/j.1469-7610.2006.01611.x.

Bleiberg, E. (2013). Mentalizing-Based Treatment with Adolescents and Families. *Child and Adolescent Psychiatry Clinics of North America*, *22*, 295–330. https://doi.org/10.1016/j.chc.2013.01.001.

Blos, P. (1967). The second individuation process of adolescence. *The Psychoanalytic Study of the Child,* 22, 162-86. https://doi.org/10.1080/00797308.1967.11822595.

Bongers, I. L., Koot, H. M., Ende, J. Van Der, & Verhulst, F. C. (2003). The Normative Development of Child and Adolescent Problem Behavior. *Journal of Abnormal Psychology*, *112*(2), 179–192. https://doi.org/10.1037/0021-843X.112.2.179.

Bouchard, M.A. & Lecours, S. (2008) Contemporary approaches to mentalization in the light of Freud's project. In Busch, F. *Mentalization: Theoretical Considerations, Research Findings, and Clinical Implications,* Chap 3, Ed. Psychoanalytic Inquiry Book Series.

Bowlby, J. (1969). *Attachment and Loss: Volume 1: Attachment*. The International Psycho- Analytical Library, London: The Hogarth Press and the Institute of Psycho-Analysis.

Bowlby, J. (1973). *Attachment and loss: Volume 2: Separation: Anxiety and anger*. New York: Basic Books. Penguin Books, 1975.

Bowlby, J. (2008). *Attachment*. Basic books.

Brandt, D. E. (1977). Separation and Identity in Adolescence—Erikson and Mahler—some.

Bremner, J. D., Randall, P., Vermetten, E., et al., (1997) Magnetic resonance imaging based measurement of hippocampal volume in posttraumatic stress disorder related to childhood physical and sexual abuse. A preliminary report. Biological Psychiatry, 41, 23-32.

Bretherton, I. (1992). The origins of attachment theory: John Bowlby and Mary Ainsworth. *Developmental Psychology, 28*(5), 759-775. http://dx.doi.org/10.1037/0012-1649.28.5.759.

Brody, A. L., Saxena, S., Schwartz, J. M., Stoessel, P. W., Maidment, K., Phelps, M. E., & Baxter Jr, L. R. (1998). FDG-PET predictors of response to behavioral therapy and pharmacotherapy in obsessive compulsive disorder. *Psychiatry Research: Neuroimaging*, *84*(1), 1-6.

Brody, A. L., Saxena, S., Stoessel, P., Gillies, L. A., Fairbanks, L. A., Alborzian, S., & Ho, M. K. (2001). Regional brain metabolic changes in patients with major depression treated with either paroxetine or interpersonal therapy: preliminary findings. *Archives of general psychiatry*, *58*(7), 631-640.

Buchheim, A., Heinrichs, M., George, C., Pokorny, D., Koops, E., Henningsen, P., O'Connor, M. & Gundel, H. (2009). Oxytocin enhances the experience of attachment security. *Psycho-neuroendocrinology, 34*(9), 1417-1422. https://doi.org/10.1016/ j.psyneuen.2009.04.002.

Bush, G., Luu, P., & Posner, M. I. (2000). Cognitive and emotional influences in anterior cingulate cortex. *Trends in cognitive sciences*, *4*(6), 215-222.

Cardinal, R., Pennicott, D., Sugathapala, L., Robbins, T. & Everitt, B. (2001). Impulsive Choice Induced in Rats by Lesions of the Nucleus Accumbens Core. *Science, 292*(5526), 2499-2501. doi: 10.1126/ science.1060818.

Carlson, N. R. (2012). *Physiology of behavior*. Pearson Higher Ed.

Carpenter, P. A., Just, M. A., & Reichle, E. D. (2000). Working memory and executive function: Evidence from neuroimaging. *Current opinion in neurobiology*, *10*(2), 195-199.

Carr, J. H. (2011). *Neurological Rehabilitation, 2/e*. Elsevier India.

Carr, L., Iacoboni, M., Dubeau, M. C., Mazziotta, J. C., & Lenzi, G. L. (2003). Neural mechanisms of empathy in humans: a relay from neural systems for imitation to limbic areas. *Proceedings of the national Academy of Sciences*, *100*(9), 5497-5502.

Carrillo-Roa, T., McGrath, C. L., Zannas, A. S., Müller-Myhsok, B., Kelley, M. E., Craighead, W. E., & Binder, E. B. (2015). Polygenic Predictors of Antidepressant Treatment-Specific Response.

Casey, B. J., Jones, R. M., & Somerville, L. H. (2011). Braking and Accelerating of the Adolescent Brain. *Journal of Research on*

Adolescence, *21*(1), 21–33. https://doi.org/10.1111/j.1532-7795.2010.00712.x.

Casey, B. J., Jones, R. & Hare, T. (2008). The adolescent brain. *Annals of the New York Academy of Sciences, 1124*(1), 111-126. https://doi.org/10.1196/annals.1440.010

Chambers, C. D., Bellgrove, M. A., Gould, I. C., English, T., Garavan, H., McNaught, E., & Mattingley, J. B. (2007). Dissociable mechanisms of cognitive control in prefrontal and premotor cortex. *Journal of neurophysiology*, *98*(6), 3638-3647.

Chiron, C., Jambaque, I., Nabbout, R., Lounes, R., Syrota, A., & Dulac, O. (1997). The right brain hemisphere is dominant in human infants. *Brain: a journal of neurology*, *120*(6), 1057-1065.

Choi-Kain, L. W., & Gunderson, J. G. (2008). The Treatment of Borderline Personality Disorder. *American Journal of Psychiatry*, *165*, 1127–1135.

Chugani, H. T. (1998). Biological basis of emotions: Brain systems and brain development. *Pediatrics*, *102*(Supplement E1), 1225-1229.

Ciarrochi, J., Bilich, L., & Godsell, C. (2010). Psychological flexibility as a mechanism of change in acceptance and commitment therapy. *Assessing mindfulness and acceptance processes in clients: Illuminating the theory and practice of change*, 51-75.

Clark, C., Prior, M., & Kinsella, G. (2002). The relationship between executive function abilities, adaptive behaviour, and academic achievement in children with externalising behaviour problems. *Journal of Child Psychology and Psychiatry*, *43*(6), 785-796.

Cloninger, C. R., Svrakic, D. M. & Pryzbeck, T. R. (1993) A psychobiological model of temperament and character. Archives of General Psychiatry, 50, 975-990.

Cohrs, R. J., Martin, T., Ghahramani, P., Bidaut, L., Higgins, P. J., & Shahzad, A. (2015). Translational medicine definition by the European Society for Translational Medicine.

Collins, A. (2003). More than Myth: The Developmental Significance of Romantic Relationships during Adolescence. *Journal of Research on Adolescence, 13*(1), 1-24. https://doi.org/10.1111/1532-7795.1301001.

Collins, A., & Laursen, B. (2004). Changing Relationships, Changing Youth: Interpersonal Contexts of Adolescent Development. *Journal of*

Early Adolescence, *24*(1), 55–62. https://doi.org/10.1177/0272431603260882.

Connolly, J., Craig, W., Golberg, A. & Pepler, D. (1999). Conceptions of Cross-Sex Friendships and Romantic Relationships in Early Adolescence. *Journal of Youth and Adolescence, 28*(4), 481-494. https://doi.org/10.1023/A:1021669024820.

Connolly, J., Pepler, D., Craig, W. & Taradash, A. (2000). Dating Experiences of Bullies in Early Adolescence. *Child Maltreatment, 5*(4), 299-310. https://doi.org/10.1177/1077559500005004002.

Conte, C. (2009). *Advanced techniques for counseling and psychotherapy*. Springer Publishing Company.

Cooper, A. & Redfern, S. (2016). *Reflective Parenting: A Guide to Understanding What's Going On in Your Child's Mind*. London: Routledge

Corbetta, M. (1998). Frontoparietal cortical networks for directing attention and the eye to visual locations: Identical, independent, or overlapping neural systems? *Proceedings of the National Academy of Sciences*, *95*(3), 831-838.

Cowan, N. (2017). The many faces of working memory and short-term storage. *Psychonomic bulletin & review*, *24*(4), 1158-1170.

Crews, F., He, J. & Hodge, C. (2007). Adolescent cortical development: A critical period of vulnerability for addiction. *Pharmacology Biochemistry and Behavior, 86*(2), 189-199. https://doi.org/10.1016/j.pbb.2006.12.001.

Critchley, H., Daly, E., Bullmore, E., Williams, S., Van Amelsvoort, T., Robertson, D., Rowe, A., Phillips, M., McAlonan, G., Howlin, P. & Murphy, D. (2000). The functional neuroanatomy of social behaviour: Changes in cerebral blood flow when people with autistic disorder process facial expressions. *Brain, A Journal of Neurology, 123*(11), 2203-2212. https://doi.org/10.1093/brain/123.11.2203.

Crone, E. A., & Dahl, R. E. (2012). Understanding adolescence as a period of social – affective engagement and goal flexibility. *Nature Reviews. Neuroscience*, *13*(9), 636=650. https://doi.org/10.1038/nrn3313.

Dahl, R. (2004). Adolescent brain development: a period of vulnerabilities and opportunities. *Annuals of New York Academy of Sciences*, *1021*, 1–11. https://doi.org/10.1196/annals.1308.001.

Damasio, A. R., & Damasio, H. (1994). Cortical systems for retrieval of concrete knowledge: The convergence zone framework. *Large-scale neuronal theories of the brain*, 6174.

Daubney, M., & Bateman, A. (2015). Mentalization-based therapy (MBT): an overview. *Australasian Psychiatry*, *23*(2), 132–135. https://doi.org/10.1177/1039856214566830.

Davey, Graham C. L., ed. *Phobias: A Handbook of Theory, Research and Treatment.* Chichester, U.K.: Wiley, 1997.

Davidov, M., & Grusec, J. E. (2006). Untangling the links of parental responsiveness to distress and warmth to child outcomes. *Child Development*, 77, 44–58.

De Rosnay, M. & Harris, P. L. (2002). Individual differences in children's understanding of emotion: The roles of attachment and language. *Attachment & Human Development*, 4, 39-54. https://doi.org/10.1080/14616730210123139.

Denham, S., Zoller, D. & Couchoud, E. (1994). Socialization of preschoolers' emotion understanding. *Developmental Psychology, 30*(6), 928-936.

Dennett, D. C., (1987) Three Kinds of Intentional Psychology, pp. 43-68 in Dennett, D. C., *The Intentional Stance,* MIT Press, (Cambridge), 1987.

Deouell, L. Y., Sacher, Y., & Soroker, N. (2005). Assessment of spatial attention after brain damage with a dynamic reaction time test. *Journal of the International Neuropsychological Society*, *11*(6), 697-707.

Deutsch, H. (1944) *The Psychology of Women,* Volume. I New York: Grune & Stratton.

Diamond, A. (2005). Attention-deficit disorder (attention-deficit/hyperactivity disorder without hyperactivity): a neurobiologically and behaviorally distinct disorder from attention-deficit/hyperactivity disorder (with hyperactivity). *Development and psychopathology*, *17*(3), 807-825.

Diamond, A. (2013). Executive functions. *Annual review of psychology*, *64*, 135-168.

Diamond, A., & Jankovic, J. (2005). The effect of deep brain stimulation on quality of life in movement disorders. *Journal of Neurology, Neurosurgery & Psychiatry*, *76*(9), 1188-1193.

Dishion, T. J., & Tipsord, J. M. (2011). Peer Contagion in Child and Adolescent Social and Emotional Development. *Annual Review of Psychology*, *62*, 189–214. https://doi.org/10.1146/annurev.psych.093008.100412.

Dolcos, F., Denkova, E., & Dolcos, S. (2012). Neural correlates of emotional memories: a review of evidence from brain imaging studies. *Psychologia*, *55*(2), 80-111.

Douvan, E. & Adelson, J. (1966). *The Adolescent Experience*. New York: John Wiley & Sons, Inc.

DuBuc, T., & Martindale, M. (2009). 03-P129. *MECHANISMS OF DEVELOPMENT*, *126*, S67-S106.

Dunn, J., Brown, J., Slomkowski, C. T., & Youngblade, L. (1991). Young children's understanding of other people's feelings and beliefs: Individual differences and their antecedents. *Child Development*, 62, 1352–1366. doi: 10.2307/1130811.

Egeland, B., Carlson, E., & Sroufe, L. A. (1993). Resilience as process. *Development and psychopathology*, *5*(4), 517-528.

Eigsti, I., Zayas, V., Mischel, W., Shoda, Y., Ayduk, O., Dadlani, M., Davidson, M., Aber, L. & Casey, B. J. (2006). Predicting Cognitive Control from Preschool to Late Adolescence and Young Adulthood. *Psychological Science, 17*(6), 478-484. https://doi.org/10.1111/j.1467-9280.2006.01732.x.

Eisner, D. A. (2000). *The death of psychotherapy: From Freud to alien abductions*. Greenwood Publishing Group.

Ellenberger, H. F. (2001). *Histoire de la découverte de l'inconscient*. Fayard.

Erikson, E & Erikson, J. M. (1997) *The Life Cycle Completed: Extended Version*. Ed. W. W. Norton & Company, New York & London.

Erikson, E H. (1970). Reflections on the dissent of contemporary youth. *The International Journal of Psychoanalysis, 51*(1), 11-22.

Erikson, E. H. (1956). The Problem of Ego Identity. *Journal of the American Psychoanalytic Association, 4*(1), 56-121. https://doi.org/ 10.1177/0003 06515600400104.

Eysenck, H. J. (1994). The outcome problem in psychotherapy: What have we learned? *Behaviour Research and Therapy*, *32*(5), 477-495.

Farah, M. J., Rabinowitz, C., Quinn, G. E., & Liu, G. T. (2000). Early commitment of neural substrates for face recognition. *Cognitive neuropsychology*, *17*(1-3), 117-123.

Fawzy, F. I., Cousins, N., Fawzy, N. W., et al., (1993) A structured psychiatric innovation for cancer patients: I: Changes over time and methods of coping and affective disturbances. Archives of General Psychiatry, 47, 720-725.

Feldman, S., Gowen, K. & Fisher, L. (1998). Family Relationships and Gender as Predictors of Romantic Intimacy in Young Adults: A Longitudinal Study. *Journal of Research on Adolescence, 8*(2), 263-286. doi: 10.1207/s15327795jra0802_5.

Fischer-Kern, M., Tmej, A., Kapusta, N., Naderer, A., Leithner-Dziubas, K., Loffler-Strstka, H. & Springer-Kremser, M. (2008). The capacity for mentalization in depressive patients: a pilot study. *Psychosomatische Medizin und Psychotherapie*, *54*(4), 368-380. doi: 10.13109/zptm. 2008.54.4.368.

Fonagy, P & Levinson, A. (2004). Offending and attachment: The relationship between interpersonal awareness and offending in a prison population with psychiatric disorder. *Canadian Journal of Psychoanalysis, 12*(2) 225-251.

Fonagy, P, & Target, M. (1997). Attachment and reflective function: their role in selforganization. *Development and Psychopathology, 9*(4), 679-700.

Fonagy, P, & Target, M. (2005). Bridging the transmission gap: an end to an important mystery. *Attachment and Human Development*, 7(3), 333-343. doi: 10.1080/14616730500269278.

Fonagy, P, Target, M, Steele, H., & Steele, M. (1998). *Reflective Functioning Manual Version 5, for Application to Adult Attachment Interviews*. London: University College London.

Fonagy, P. & Luyten, P. (2009) A developmental mentalization-based approach to the understanding and treatment of borderline personality disorder. *Development and Psychopathology, 21,* 1355-1381. doi: 10.10 17/S0954579409990198.

Fonagy, P. (1991). Thinking about thinking: some clinical and theoretical considerations in the treatment of a borderline patient. *The International Journal of Psychoanalysis, 72*(4), 639-656.

Fonagy, P. (1999). Points of contact and divergence between psychoanalytic and attachment theories: Is psychoanalytic theory truly different. *Psychoanalytic Inquiry, 19*, 448-480. http://dx.doi.org/ 10.1080/07351 699909534264.

Fonagy, P. (2008). The mentalization-focused approach in social development. In Busch, F. *Mentalization: Theoretical Considerations, Research Findings, and Clinical Implications*, Chapter 1, Ed. Psychoanalytic Inquiry Book Series.

Fonagy, P., & Allison, E. (2014). The Role of Mentalizing and Epistemic Trust in the Therapeutic Relationship. *Psychotherapy*, *51*(3), 372–380. https://doi.org/10.1037/a0036505.

Fonagy, P., & Bateman, A. W. (2006). Mechanisms of Change in Mentalization-Based Treatment of BPD. *Journal of Clinical Psychology*, *62*(4), 411–430. https://doi.org/10.1002/jclp.

Fonagy, P., & Target, M. (1997). Attachment and reflective function: Their role in self-organization. *Development and psychopathology*, *9*(4), 679-700.

Fonagy, P., & Target, M. (1997). Attachment and reflective function: their role in self-organization. *Development and Psychopathology*, *9*(4), 679–700.

Fonagy, P., Bateman, A. & Luyten, P. (2012) Introduction and overview. In Bateman, A. & Fonagy, P. *Handbook of mentalizing in mental health practice*, Chapter 1, Ed. American Psychiatric Association.

Fonagy, P., Bateman, A., & Bateman, A. (2011). The widening scope of mentalizing: A discussion. *Psychology and Psychotherapy: Theory, Research and Practice*, *84*, 98–110. https://doi.org/10.1111/j.2044-8341.2010.02005.x.

Fonagy, P., Gergely, G. & Target, M. (2007). The parent–infant dyad and the construction of the subjective self. *The Journal of Child Psychology and Psychiatry, 48*(3-4), 288-328. https://doi.org/10.1111/j.1469-7610.2007.01727.x.

Fonagy, P., Gergely, G., Jurist, E. & Target, M. (2003) *Affect Regulation, Mentalization and the Development of the Self*, Ed. Karnac Books

Fonagy, P., Rossouw, T., Sharp, C., Bateman, A., Allison, L., & Farrar, C. (2014). Mentalization-Based Treatment for Adolescents with Borderline Traits. In C. Sharp & J. L. Tackett (Eds.), *Handbook of Borderline Personality Disorder in Children and Adolescents* (pp. 313–332). Springer. https://doi.org/10.1007/978-1-4939-0591-1.

Fox, K. C., Nijeboer, S., Dixon, M. L., Floman, J. L., Ellamil, M., Rumak, S. P., & Christoff, K. (2014). Is meditation associated with altered brain structure? A systematic review and meta-analysis of morphometric neuroimaging in meditation practitioners. *Neuroscience & Biobehavioral Reviews*, *43*, 48-73.

Freeman, C. (2016). What is mentalizing? An overview. *British Journal of Psychotherapy*, *32*(2), 189–201. https://doi.org/10.1111/bjp.12220.

Freud, A. (1958). Adolescence. *Psychoanalytic Study of the Child, 13*, 255-278.

Freud, S. (1905). *Three Essays on the Theory of Sexuality. The Standard Edition of the Complete Psychological Works of Sigmund Freud*, Volume VII (1901-1905): A Case of Hysteria, Three Essays on Sexuality and Other Works (pp. 123-246).

Freud, S. (1911). *Formulations on the Two Principles of Mental Functioning. The Standard Edition of the Complete Psychological Works of Sigmund Freud*, Volume XII (1911-1913): The Case of Schreber, Papers on Technique and Other Works (pp. 213-226).

Freud, S. (1914) Remembering, repeating, and working-through (Further Recommendations on the Technique of Psycho-analysis II). Reprinted (1953-1974) in the Standard Edition of the Complete Psychological Works of Sigmund Freud (trans. & ed. J. Strachey), vol. 12, pp. 145-156. London: Hogarth Press.

Freud, S. (1914). *On Narcissism: An Introduction. The Standard Edition of the Complete Psychological Works of Sigmund Freud*, Volume XIV

(1914-1916): On the History of the Psycho-Analytic Movement, Papers on Metapsychology and Other Works, 67-102

Frewen, P. A., Dozois, D. J., & Lanius, R. A. (2010). Neuroimaging studies of psychological interventions for mood and anxiety disorders: empirical and methodological review. *Focus*, *8*(1), 92-109.

Fries, A., Ziegler, T., Kurian, J., Jacoris, S. & Pollak, S. (2005). Early experience in humans is associated with changes in neuropeptides critical for regulating social behavior. *Proceedings of the National Academy of Science of the United States of America, 102*(47), 17237-17240. https://doi.org/10.1073/pnas.0504767102.

Frith, C. & Frith, U. (2006). The Neural Basis of Mentalizing. *Neuron, 50*(4), 531-534. https://doi.org/10.1016/j.neuron.2006.05.001.

Frith, C. D., & Frith, U. (2012). Mechanisms of social cognition. *Annual review of psychology*, *63*, 287-313.

Frith, U. & Frith, C. (2003). Development and neurophysiology of mentalizing. *Philosophical Transactions of the Royal Society of London. Series B: Biological Sciences*, 358(1431), 459-473. https://doi.org/10.1098/rstb.2002.1218.

Frith, U., & Frith, C. (2001). The biological basis of social interaction. *Current Directions in Psychological Science*, *10*(5), 151-155.

Fu, C. H., Steiner, H., & Costafreda, S. G. (2013). Predictive neural biomarkers of clinical response in depression: a meta-analysis of functional and structural neuroimaging studies of pharmacological and psychological therapies. *Neurobiology of disease*, *52*, 75-83.

Furman, W. & Buhmester, D. (1992). Age and sex differences in perceptions of networks and personal relationships. *Child Development*, *63*, 103–115. https://doi.org/10.1111/j.1467-8624.1992.tb03599.x.

Furmark. T., Tillfors, M., Marteinsdottir, I., Fischer, H., Pissiota, A., Landstrom, B. & Fredrikson, M. (2002). Common Changes in Cerebral Blood Flow in Patients With Social Phobia Treated With Citalopram or Cognitive-Behavioral Therapy. *Archives of General Psychiatry, 59*(5), 425–433. doi:10.1001/archpsyc.59.5.425.

Fuster, J. (2015). *The prefrontal cortex*. Academic Press.

Fuster, J. M. (2001). The prefrontal cortex—an update: time is of the essence. *Neuron*, *30*(2), 319-333.

Gabbard, G. O. (1994). Mind and brain in psychiatric treatment. *Bulletin of the Menninger Clinic*, *58*(4), 427.

Gabbard, G. O. (2000). A neurobiologically informed perspective on psychotherapy. *The British Journal of Psychiatry*, *177*(2), 117-122.

Gardner, M. & Steinberg, L. (2005). Peer influence on risk taking, risk preference, and risky decision making in adolescence and adulthood: an experimental study. *Developmental Psychology, 41*(4), 625-635. doi: 10.1037/0012-1649.41.4.625.

Gazzaley, A., Rissman, J., & D'esposito, M. (2004). Functional connectivity during working memory maintenance. *Cognitive, Affective, & Behavioral Neuroscience*, *4*(4), 580-599.

Gazzaniga, M. S. (2005). Forty-five years of split-brain research and still going strong. *Nature Reviews Neuroscience*, *6*(8), 653.

Gergely, G. & Unoka, Z. (2008). The development of the unreflective self. In Busch, F. *Mentalization: Theoretical Considerations, Research Findings, and Clinical Implications,* Chapter 2, Ed. Psychoanalytic Inquiry Book Series.

Gergely, G. & Watson, J. (1996). The social biofeedback model of parental affect-mirroring. *The International Journal of Psychoanalysis, 77*(6), 1181-1212.

Giedd, J. N. (2008). The teen brain: insights from neuroimaging. *Journal of Adolescent Health*, *42*(4), 335-343.

Giedd, J. N., & Rapoport, J. L. (2010). Structural MRI of pediatric brain development: what have we learned and where are we going? *Neuron*, *67*(5), 728-734.

Gilbert, S., Spengler, S., Simons, J. & Steele, D. (2006). Functional Specialization within Rostral Prefrontal Cortex (Area 10): A Meta-analysis. *Journal of Cognitive Neuroscience, 18*(6), 932-948. https://doi.org/10.1162/jocn.2006.18.6.932.

Giovacchini, P. L. (1973). Character disorders: With special reference to the borderline state. *International Journal of Psychoanalytic Psychotherapy*.

Goldapple, K., Segal, Z., Garson, C., Lau, M., Bieling, P., Kennedy, S., & Mayberg, H. (2004). Modulation of cortical-limbic pathways in major depression: treatment-specific effects of cognitive behavior therapy. *Archives of general psychiatry*, *61*(1), 34-41.

Goldman, A. I. (2012). Theory of mind. *The Oxford handbook of philosophy of cognitive science*, 402-424.

Goodwin, S. J. (2018). Neurogenesis: remembering all or forgetting some. *Journal of neurophysiology*.

Gorman, J. & Roose, S. (2011). The Neurobiology of Fear Memory Reconsolidation and Psychoanalytic Theory. *Journal of the American Psychoanalytic Association, 59*(6), 1201-1220. https://doi.org/10.1177/0003065111427724.

Gottman, J. M., Katz, L. F., & Hooven, C. (1997). Meta-emotion. Hillsdale, NJ: Erlbaum. Kagitcibasi, C. (1970). Social norms and authoritarianism: A Turkish–American comparison. *Journal of Personality and Social Psychology, 16*, 444-451.

Gourevitch, A. (1980). Three Aspects of Change in Adolescence. *Contemporary Psychoanalysis, 16*(4), 547-564. https://doi.org/10.1080/00107530.1980.10745637.

Grawe, K. (2017). *Neuropsychotherapy: How the neurosciences inform effective psychotherapy*. Routledge.

Greene, J. & Haidt, J. (2002). How (and where) does moral judgment work? *Trends in Cognitive Sciences, 6*(12), 517-523. https://doi.org/10.1016/S1364-6613(02)02011-9.

Greenough, W. T., Black, J. E & Wallace, C. S. (1987) Experience and brain development. Child Development, 58, 539-559.

Grienenberger, J. F., Kelly, K., & Slade, A. (2005). Maternal reflective functioning, mother infant affective communication, and infant attachment: exploring the link between mental states and observed caregiving behavior in the intergenerational transmission of attachment. *Attachment & human development*, *7*(3), 299-311. doi: 10.1080/14616730500245963.

Grosjean, B. (2005). From synapse to psychotherapy: the fascinating evolution of neuroscience. *American Journal of psychotherapy*, *59*(3), 181-197.

Gross, J. J. (2014a). Emotion regulation: Conceptual and empirical foundations. In J. J. Gross (Ed.), *Handbook of emotion regulation* (2nd ed., pp. 3–20). New York, NY: Guilford.

Gross, J. J. (1998). Antecedent- and response-focused emotion regulation: Divergent consequences for experience, expression, and physiology. *Journal of Personality and Social Psychology, 74,* 224–237

Gunnar, M., & Quevedo, K. (2007). The neurobiology of stress and development. *Annu. Rev. Psychol.*, *58*, 145-173.

Gunnar, M., & Quevedo, K. (2007). The neurobiology of stress and development. *Annu. Rev. Psychol.*, *58*, 145-173.

Gunther Moor, B., van Leijenhorst, L., Rombouts, S. A., Crone E. A. & Van der Molenref, M. W. (2010). Do you like me? Neural correlates of social evaluation and developmental trajectories. *Social Neuroscience, 5*(5-6), 461-482. doi: 10.1080/17470910903526155.

Gusnard, D. A, Akbudak, E., Shulman, G. L. & Raichle, M. E. (2001). Medial prefrontal cortex and self-referential mental activity: relation to a default mode of brain function. *Proceedings of the National Academy of Sciences of the United States of America, 98*(7), 4259-4264.

Gweon, H., Dodell-Feder, D., Bedny, M & Saxe, R. (2012). Theory of mind performance in children correlates with functional specialization of a brain region for thinking about thoughts. *Child Development, 83*(6), 1853-1868.

Haramati, S., Soroker, N., Dudai, Y., & Levy, D. A. (2008). The posterior parietal cortex in recognition memory: a neuropsychological study. *Neuropsychologia*, *46*(7), 1756-1766.

Haris, P. L. & Gross, D. (1988). Children's Understanding of Real and Apparent Emotion. In

Hartlage, L. C., & D'Amato, R. C. (2008). Understanding the etiology of psychiatric and neurologic disorders in neuropsychiatry. *The neuropsychology handbook*, 87-108.

Hayes, S. C., Luoma, J. B., Bond, F. W., Masuda, A., & Lillis, J. (2006). Acceptance and commitment therapy: Model, processes and outcomes. *Behaviour research and therapy*, *44*(1), 1-25.

Hecht, D. (2014). Cerebral lateralization of pro-and anti-social tendencies. *Experimental neurobiology*, *23*(1), 1-27.

Heim, C., Newport, J., Mletzko, T., Miller, A. & Nemeroff, C. (2008). The link between childhood trauma and depression: Insights from HPA axis studies in humans. *Psychoneuroendocrinology, 33*(6), 693-710. https://doi.org/10.1016/j.psyneuen.2008.03.008.

Hermoye, L., Saint-Martin, C., Cosnard, G., Lee, S. K., Kim, J., Nassogne, M. C., & Wakana, S. (2006). Pediatric diffusion tensor imaging: normal database and observation of the white matter maturation in early childhood. *Neuroimage*, *29*(2), 493-504.).

HESS, W. R., & AKERT, K. (1955). Experimental data on role of hypothalamus in mechanism of emotional behavior. *AMA Archives of Neurology & Psychiatry*, *73*(2), 127-129.

Hollander, E. (1997). *Obsessive-compulsive disorders: diagnosis, etiology, treatment*. Marcel Dekker.

Holmes, J. (2006) Mentalizing from a psychoanalytic perspective: what's new? In. Allen, J. & Fonagy, P. *Handbook of mentalization-based treatment*, Chapter 2, Ed. Wiley & Sons Ltd

Hopkins, J. (1992). Psychoanalysis, interpretation, and science. In J. Hopkins & A. Savile (Eds.), *Psychoanalysis, Mind and Art: Perspectives on Richard Wollheim* (pp. 3–34). Oxford: Blackwell.

Hughes, C., & Dunn, J. (1998). Understanding mind and emotion: Longitudinal associations with mental-state talk between young friends. *Developmental Psychology, 34*(5), 1026-1037. http://dx.doi. org/10.10 37/0012-1649.34.5.1026.

Iacoboni, M., & Dapretto, M. (2006). The mirror neuron system and the consequences of its dysfunction. *Nature Reviews Neuroscience*, *7*(12), 942.

Iacoboni, M., Woods, R. P., Brass, M., Bekkering, H., Mazziotta, J. C., & Rizzolatti, G. (1999). Cortical mechanisms of human imitation. *science*, *286*(5449), 2526-2528.

Insel, T. & Young, L. (2000). Neuropeptides and the evolution of social behavior. *Current Opinion in Neurobiology, 10*(6), 784-789. https://doi.org/10.1016/S0959-4388(00)00146-X.

Insel, T. (2003). Is social attachment an addictive disorder? *Physiology & Behavior, 79*(3), 351-357. https://doi.org/10.1016/S0031-9384(03) 00148-3.

Insel, T. R., & Cuthbert, B. N. (2015). Brain disorders? Precisely. *Science*, *348*(6234), 499-500.

Insel, T. R., & Fenton, W. S. (2005). Psychiatric epidemiology: it's not just about counting anymore. *Archives of general psychiatry*, *62*(6), 590-592.).

Jacobson, E. (1961). Adolescent Moods and the Remodeling of Psychic Structures in Adolescence. *Psychoanalytic Study of the Child, 16*, 164-183. https://doi.org/10.1080/00797308.1961.11823204.

Jay, N. (2003). The anatomy of mentalization : A view from developmental neuroimaging.

Jenkins, J. M. & Astington, J. W. (1996). Cognitive factors and family structure associated with theory of mind development in young children. *Developmental Psychology, 32*, 70–78. http://dx.doi.org/10.1037/0012-1649.32.1.70.

Joffe, R., Segal, Z. & Singer, W. (1996) Change in thyroid hormone levels following response to cognitive therapy for major depression. American Journal of Psychiatry, 153, 411-413.

Johnson, M. H., & Karmiloff-Smith, A. (2004). Neuroscience perspectives on infant development. *Theories of infant development*, 121-141.

Jurist, E. (2010). Mentalizing Minds. *Psychoanalytic Inquiry*, *30*(4), 289–300. https://doi.org/10.1080/07351690903206496.

Jurist, E. L., & Meehan, K. B. (2009). Attachment, mentalization, and reflective functioning. In J. H. Obegi & E. Berant (Eds.), *Attachment theory and research in clinical work with adults* (pp. 71-93). New York, NY, US: Guilford Press.

Kandel, D. & Lesser, G. (1972). Youth on Two Worlds: United States and *Denmark*. San Fransisco: Jossey-Bass.

Kandel, E. R. (1998) A new intellectual framework for psychiatry. American Journal of Psychiatry, 155, 457-469.

Karlsson, H. (2011). How psychotherapy changes the brain. *Psychiatric Times*, *28*(8), 0.

Karlsson, H. (2011). How psychotherapy changes the brain: Understanding the mechanisms. *Psychiatric Times*, *28*(8), 21-21.

Kashdan, T. B., & Rottenberg, J. (2010). Psychological flexibility as a fundamental aspect of health. *Clinical psychology review*, *30*(7), 865-878.

Kashdan, T. B., Barrios, V., Forsyth, J. P., & Steger, M. F. (2006). Experiential avoidance as a generalized psychological vulnerability: Comparisons with coping and emotion regulation strategies. *Behaviour research and therapy*, *44*(9), 1301-1320.

Katz, D. H. (2014). *Lymphocyte differentiation, recognition, and regulation*. Academic Press.).

Katznelson, H. (2014). Reflective Functioning: A review. *Clinical Psychology Review*, *34*, 107–117. https://doi.org/10.1016/j.cpr.2013.12.003.

Kendler, K. S., Kessler, R. C., Neale, M. C., et al (1995) Stressful life events, genetic liability, and onset of major depression in women. American Journal of Psychiatry, 152, 833-842.

Kennedy, S. H., Konarski, J. Z., Segal, Z. V., Lau, M. A., Bieling, P. J., McIntyre, R. S., & Mayberg, H. S. (2007). Differences in brain glucose metabolism between responders to CBT and venlafaxine in a 16-week randomized controlled trial. *American Journal of Psychiatry*, *164*(5), 778-788.

Klimes-Dougan, B., & Zeman, J. (2007). Introduction to the special issue of social development: Emotion socialization in childhood and adolescence. *Social Development, 16,* 203–209. https://doi.org/10.1111/j.1467-9507.2007.00380.x.

Kobak, R. & Ferenz-Gillies, R. (1995). Emotion regulation and depressive symptoms during adolescence: A functionalist perspective. *Development and Psychopathology, 7*(1), 183-192. https://doi.org/10.1017/S0954579400006416.

Kobak, R., Sudler, N. & Gamble, W. (1991). Attachment and depressive symptoms during adolescence: A developmental pathways analysis. *Development and Psychopathology, 3*(4), 461-474. https://doi.org/10.1017/S095457940000763X.

Kochanska, G., & Knaack, A. (2003). Effortful control as a personality characteristic of young children: Antecedents, correlates, and consequences. *Journal of personality*, *71*(6), 1087-1112.

Kolb, B. (2013). *Brain plasticity and behavior*. Psychology Press.

KOLB, B., & FANTIE, B. (2013). Development of the Child's Brain and. *Handbook of Clinical Child Neuropsychology*, 17.

Kolb, B., & Fantie, B. D. (2009). Development of the child's brain and behavior. In *Handbook of clinical child neuropsychology* (pp. 19-46). Springer, Boston, MA.

Korb, A. (2015). *The upward spiral: Using neuroscience to reverse the course of depression, one small change at a time*. New Harbinger Publications.

Koren-Karie, N. I. N. A., Oppenheim, D., & Goldsmith, D. F. (2007). Keeping the Inner World of the Child in Mind. *Attachment Theory in Clinical Work with Children: Bridging the Gap Between Research and Practice*, 31.

Kovacs, Á. M., Téglás, E., and Endress, A. D. (2010). The social sense: susceptibility to others' beliefs in human infants and adults. *Science, 330*, 1830–1834. doi: 10.1126/science.1190792.

Kroger, J. (2006). Identity Development during Adolescence. In Adams, G. & Berzonsky, M. *Blackwell Handbook of Adolescence*, Chapter 10, Ed. Blackwell Publishing.

Kumari V., Fannon D., Peters E.R. et al. (2011) Neural changes following cognitive behaviour therapy for psychosis: a longitudinal study. Brain, 134(8): 2396–2407.

Lanius, R. A., Frewen, P. A., Tursich, M., Jetly, R., & McKinnon, M. C. (2015). Restoring large-scale brain networks in PTSD and related disorders: a proposal for neuroscientifically-informed treatment interventions. *European journal of psychotraumatology*, *6*(1), 27313.

Laplanche, J. & Pontalis. J. B. (1973). *The Language of Psychoanalysis*, Hogarth Press Ltd

Laufer, M. (1964). Ego Ideal and Pseudo Ego Ideal in Adolescence. *The Psychoanalytic Study of the Child, 19*(1), 196-221. https://doi.org/10.1080/00797308.1964.11822865.

Lecours, S., & Bouchard, M.-A. (1997). Dimensions of mentalisation: Outlining levels of psychic transformation. *The International Journal of Psychoanalysis*, *78*, 855–875.

LeDoux, J. E. (1994). Emotion, memory and the brain. *Scientific American, 270*(6), 50-57.

LeDoux, J. E., & Pine, D. S. (2016). Using neuroscience to help understand fear and anxiety: a two-system framework. *American journal of psychiatry, 173*(11), 1083-1093.

Lehto, J. E., Juujärvi, P., Kooistra, L., & Pulkkinen, L. (2003). Dimensions of executive functioning: Evidence from children. *British Journal of Developmental Psychology, 21*(1), 59-80.

Lemche, E., Giampietro, V. P., Surguladze, S. A., Amaro, E. J., Andrew, C. M., Williams, S. C., ... & Simmons, A. (2006). Human attachment security is mediated by the amygdala: Evidence from combined fMRI and psychophysiological measures. *Human brain mapping, 27*(8), 623-635.

Lemma, A. (2009). Being seen or being watched? A psychoanalytic perspective on body dysmorphia perspective on body dysmorphia. *The International Journal of Psychoanalysis, 90*(4), 753–771. https://doi.org/10.1111/j.1745-8315.2009.00158.x.

Lemma, A. (2015). Psychoanalysis in times of technoculture: Some reflections on the fate of the body in virtual space. *The International Journal of Psychoanalysis, 96*(3), 569-582. https://doi.org/10.1111/1745-8315.12348.

Levy, K. N., & Blatt, S. J. (1999). Attachment theory and psychoanalysis: Further differentiation within insecure attachment patterns. *Psychoanalytic Inquiry, 19*(4), 541-575.

Levy, R. A., Ablon, J. S., & Kächele, H. (Eds.). (2011). *Psychodynamic psychotherapy research: Evidence-based practice and practice-based evidence*. Springer Science & Business Media.

Lieberman, M. (2007). Social Cognitive Neuroscience: A Review of Core Processes. *Annual Review of Psychology, 58*, 259-289. https://doi.org/10.1146/annurev.psych.58.110405.085654.

Linden, D. E. J. (2006). How psychotherapy changes the brain–the contribution of functional neuroimaging. *Molecular psychiatry, 11*(6), 528.

Lombardo, M., Chakrabati, B., Bullmore, E. & Wheelwright, S. (2010). Shared Neural Circuits for Mentalizing about the Self and Others.

Journal of Cognitive Neuroscience, 22(7), 1623-1635. https://doi.org/10.1162/jocn.2009.21287.

Luquet, P. (1988), Langage, pensée et structure psychique. *Revue Français de Psychoanalyse, 52*, 267-302.

Luria, A. R. (2012). *Higher cortical functions in man*. Springer Science & Business Media.

Luyten, P., & Fonagy, P. (2015). The Neurobiology of Mentalizing. *Personality Disorders, Theory, Research and Treatment*, *6*(4), 366–379. Retrieved from http://dx.doi.org/10.1037/per0000117.

Luyten, P., Houdenhove, B. Van, Lemma, A., Fonagy, P., Luyten, P., Houdenhove, B. Van, & Lemma, A. (2012). A mentalization-based approach to the understanding and treatment of functional somatic disorders. *Psychoanalytic Psychotherapy*, *26*(2), 121–140. https://doi.org/10.1080/02668734.2012.678061.

Lyons-Ruth, K. & members of the Change Process Study Group (1998) Implicit relational knowing: its role in development and psychoanalytic treatment. Infant Mental Health Journal, 19, 282-289.

MacQueen, G. M., Campbell, S., McEwen, B. S., Macdonald, K., Amano, S., Joffe, R. T., & Young, L. T. (2003). Course of illness, hippocampal function, and hippocampal volume in major depression. *Proceedings of the national academy of sciences*, *100*(3), 1387-1392.

Mahler, M., (1963) Thoughts about development and individuation. *Psychoanalytic Study of the Child, 18,* 307-324. https://doi.org/10.1080/00797308.1963.11822933.

Manglik, H. The Social Brain in Action: The Human Mirror System Dynamically Expands to Include More People.

Marcia, J. E. (1966). Development and validation of ego-identity status. *Journal of Personality and Social Psychology, 3*(5), 551-558. http://dx.doi.org/10.1037/h0023281.

Martin, S. D., Martin, E., Rai, S. S., Richardson, M. A., & Royall, R. (2001). Brain blood flow changes in depressed patients treated with interpersonal psychotherapy or venlafaxine hydrochloride: preliminary findings. *Archives of General Psychiatry*, *58*(7), 641-648.

Marty, P. (1968), A major process of somatization: The progressive disorganization. *International Journal of Psychoanalysis*, *49*, 246-249.

Marty, P. (1991). *Mentalisation et Psychosomatique*. Paris: Laboratoire Delagrange.

Masten, C., Einsenberger, N., Borofsky, L., Pfeifer, J., McNealy, K., Mazziotta, J. & Dapretto, M. (2009). Neural correlates of social exclusion during adolescence: understanding the distress of peer rejection. *Social Cognitive and Affective Neuroscience, 4*(2), 143-157. https://doi.org/10.1093/scan/nsp007.

Masten, C., Telzer, E., Fuligni, A., Lieberman, M. & Eisenberger, N. (2012). Time spent with friends in adolescence relates to less neural sensitivity to later peer rejection. *Social Cognitive and Affective Neuroscience, 7*(1), 106-114. https://doi.org/10.1093/scan/nsq098.

Mayberg, H. S. (2003). Modulating dysfunctional limbic-cortical circuits in depression: towards development of brain-based algorithms for diagnosis and optimised treatment. *British medical bulletin*, *65*(1), 193-207.

Mcgauley, G., Yakeley, J., Williams, A., & Bateman, A. (2011). Attachment, mentalization and antisocial personality disorder: The possible contribution of mentalization-based treatment. *European Journal of Psychotherapy & Counselling*, *13*(4), 371–393. https://doi.org/10.1080/13642537.2011.629118.

Meins, E. (1999) Sensitivity, security and internal working models: bridging the transmission gap. *Attachment and Human Development, 1,* 325-342. doi: 10.1080/14616739900134181.

Meltzoff, A. N., & Decety, J. (2003). What imitation tells us about social cognition: a rapprochement between developmental psychology and cognitive neuroscience. *Philosophical Transactions of the Royal Society of London B: Biological Sciences*, *358*(1431), 491-500.

Mento, G., Suppiej, A., Altoè, G., & Bisiacchi, P. S. (2010). Functional hemispheric asymmetries in humans: electrophysiological evidence from preterm infants. *European Journal of Neuroscience*, *31*(3), 565-574.

Mervaala, E., Föhr, J., Könönen, M., Valkonen-Korhonen, M., Vainio, P., Partanen, K., & Lehtonen, J. (2000). Quantitative MRI of the hippocampus and amygdala in severe depression. *Psychological medicine*, *30*(1), 117-125.

Meyer-Lindenberg, A., & Tost, H. (2012). Neural mechanisms of social risk for psychiatric disorders. *Nature neuroscience*, *15*(5), 663.

Midgley, N. & Vrouva, I. (2012) *Minding the Child: Mentalization-Based Interventions with Children, Young People and their Families*, London: Routledge.

Midgley, N., Ensink, K., Lindqvist, K., Malberg, N., & Muller, N. (2017). *Mentalization-Based Treatment for Children: A Time-Limited Approach.* Washington, DC, US: American Psychological Association. http://dx.doi.org/10.1037/0000028-000.

Mischel, W., & Ebbesen, E. B. (1970). Attention in delay of gratification. *Journal of Personality and Social Psychology, 16*(2), 329-337. http://dx.doi.org/10.1037/h0029815.

Mischel, W., Ebbesen, E. B., & Raskoff Zeiss, A. (1972). Cognitive and attentional mechanisms in delay of gratification. *Journal of Personality and Social Psychology, 21*(2), 204-218. http://dx.doi.org/10.1037/h0032198.

Mischel, W., Schoda, Y. & Rodriguez, M. (1989). Delay of gratification in children. *Science, 244*(4907), 933-938. doi: 10.1126/science.2658056.

Moore, M., Culpepper, S., Phan, K. L., Strauman, T. J., Dolcos, F., & Dolcos, S. (2018). Neurobehavioral Mechanisms of Resilience Against Emotional Distress: An Integrative Brain-Personality-Symptom Approach Using Structural Equation Modeling. *Personality Neuroscience*, *1*.

Moretti, M., & Peled, M. (2004). Adolescent-parent attachment: Bonds that support healthy development. *Paediatrics Child Health*, *9*(8), 551–555.

Morris, A. S., Silk, J. S., Steinberg, L., Myers, S. S., & Robinson, L. R. (2007). The role of the family context in the development of emotion regulation. *Social Development, 16,* 361–388.

Münte, T. F., Altenmüller, E., & Jäncke, L. (2002). The musician's brain as a model of neuroplasticity. *Nature Reviews Neuroscience*, *3*(6), 473.

Nadeau, M. E., Nolin, P., & Chartrand, C. (2013). Behavioral and emotional profiles of neglected children. *Journal of Child & Adolescent Trauma*, *6*(1), 11-24.

Nelson, E., Leibenluft, E., McClure, E., & Pine, D. (2005). The social re-orientation of adolescence: a neuroscience perspective on the process and its relation to psychopathology. *Psychological Medicine*, *35*, 163–174. https://doi.org/10.1017/S0033291704003915.

Nitschke, J., Nelson, E., Rusch, B., Fox, A., Oakes, T. & Davidson, R. (2004). Orbitofrontal cortex tracks positive mood in mothers viewing pictures of their newborn infants. *NeuroImage, 21*(2), 583-592. https://doi.org/10.1016/j.neuroimage.2003.10.005.

O'Neal, C. R., & Magai, C. (2005). Do parents respond in different ways when children feel different emotions? The emotional context of parenting. *Development and Psychopathology, 17,* 467–487. https://doi.org/10.1017/S0954579405050224.

Ochsner, K. N., & Gross, J. J. (2004). Thinking makes it so: A social cognitive neuroscience approach to emotion regulation. In R. F. Baumeister & K. D. Vohs (Eds.), *Handbook of Self-Regulation: Research, Theory, and Applications* (pp. 229-255). New York, NY, US: Guilford Press.

Ochsner, K. N., & Gross, J. J. (2005). The cognitive control of emotion. *Trends in cognitive sciences*, *9*(5), 242-249.

Ochsner, K. N., Ray, R. D., Cooper, J. C., Robertson, E. R., Chopra, S., Gabrieli, J. D., & Gross, J. J. (2004). For better or for worse: neural systems supporting the cognitive down-and up-regulation of negative emotion. *Neuroimage*, *23*(2), 483-499.

Ollech, D., & McCarthy, J. (1997). Impediments to identity formation in female adolescents. *Psychoanalytic Psychology, 14*(1), 65-80. http://dx.doi.org/10.1037/h0079724.

Ornitz, E. M. (1996). Developmental aspects of neurophysiology. *Child and Adolescent Psychiatry: A Comprehensive Textbook 2nd ed. Baltimore: Williams and Wilkins*, 39-51.

Parent, A.S., Teilmann, G., Juul, A., Skakkebaek, N., Toppari, J. & Bouguignon, J.P. (2003). The Timing of Normal Puberty and the Age Limits of Sexual Precocity: Variations around the World, Secular Trends, and Changes after Migration. *Endocrine Reviews*, *24*(5), 668–693, https://doi.org/10.1210/er.2002-0019.

Patterson, C. L., Uhlin, B., & Anderson, T. (2008). Clients' pretreatment counseling expectations as predictors of the working alliance. *Journal of Counseling Psychology*, *55*(4), 528.

Patton, G., Coffrey, C., Carlin, J., Degenhardt, L., Lynksey, M. & Hall, W. (2004). Cannabis use and mental health in young people: cohort study. *BMJ, 325,* 1195-1198.

Paus, T., Keshavan, M., & Giedd, J. (2008). Why do many psychiatric disorder emerge during adolescence? *Nature Reviews. Neuroscience*, *9*, 947–957.

Perry, B. D., Pollard, R. A., Blakley, T. L., Baker, W. L., & Vigilante, D. (1995). Childhood trauma, the neurobiology of adaptation, and "use-dependent" development of the brain: How "states" become "traits". *Infant mental health journal*, *16*(4), 271-291.

Pezawas, L. (2015). Serotonin Transporter Linked Polymorphic Region: From Behavior to Neural Mechanisms. *Biological psychiatry*, *78*(8), 522-524.

Phares, V., Renk, K., Duhig, A., Fields S. & Sly, J. (2009). Gender Differences in Positive and Negative Feelings Between Adolescents and Their Fathers and Mothers. *Journal of Child and Family Studies, 18*(2), 213-218. https://doi.org/10.1007/s10826-008-9221-2.

Pinaud, R. (2004). Experience-dependent immediate early gene expression in the adult central nervous system: evidence from enriched-environment studies. *International Journal of Neuroscience*, *114*(3), 321-333.

Posner, M. I., & Fan, J. (2008). Attention as an organ system. *Topics in integrative neuroscience*, 31-61.

Posner, M. I., & Petersen, S. E. (1990). The attention system of the human brain. *Annual review of neuroscience*, *13*(1), 25-42.

Premack, D. & Woodruff, G. (1978). Does the chimpanzee have a "theory of mind"? *Behavior and Brain Sciences, 4,* 515-526. https://doi.org/10.1017/S0140525X00076512.

Primus, R. & Kellogg, C. (1989). Pubertal-related changes influence the development of environment-related social interaction in the male rat. *Developmental Psychobiology, 22*(6), 633-643. https://doi.org/10.1002/dev.420220608.

Ptak, R. (2012). The front parietal attention network of the human brain: action, saliency, and a priority map of the environment. *The Neuroscientist*, *18*(5), 502-515.

Ptak, R., & Schnider, A. (2011). The attention network of the human brain: relating structural damage associated with spatial neglect to functional imaging correlates of spatial attention. *Neuropsychologia*, *49*(11), 3063-3070.

Putnam, F. W. & Trickett, P. K. (1997) *Psychobiological effects of sexual abuse: a longitudinal study.* In Psychobiology of Posttraumatic Stress Disorder (eds R. Yehuda & A. C. McFarland), pp. 150-159. New York: New York Academy of Sciences.

Pynoos, R. S., Steinberg, A. M., & Aronson, L. (1997). Traumatic experiences: The early organization of memory in school-age children and adolescents. *Trauma and memory: Clinical and legal controversies*, 272-289.

Reiss, D., Hetherington, E. M., Plomin, R., et al., (1995) *Genetic questions for environmental studies: differential parenting and psychopathology in adolescence*. Archives of General Psychiatry, 52, 925-936.

Reyna, V. & Farley, F. (2006). Risk and Rationality in Adolescent Decision Making: Implications for Theory, Practice, and Public Policy. *Psychological Science in the Public Interest, 7*(1), 1-44. https://doi.org/10.1111/j.1529-1006.2006.00026.x.

Richardson (Eds.), *Attachment in Middle Childhood* (pp. 137–160). New York: Guilford Press.

Ripoll, L., Snyder, R., Steele, H. & Siever, L. (2013). The Neurobiology of Empathy in Borderline Personality Disorder. *Current Psychiatry Reports, 15*, 344-354. https://doi.org/10.1007/s11920-012-0344-1.

Rissman, J., & Wagner, A. D. (2012). Distributed representations in memory: insights from functional brain imaging. *Annual review of psychology*, *63*, 101-128.

Rissman, J., Greely, H. T., & Wagner, A. D. (2010). Detecting individual memories through the neural decoding of memory states and past experience. *Proceedings of the National Academy of Sciences*, *107*(21), 9849-9854.

Rizzolatti, G., & Sinigaglia, C. (2008). *Mirrors in the brain: How our minds share actions and emotions*. Oxford University Press, USA.

Rizzolatti, G., & Sinigaglia, C. (2016). The mirror mechanism: a basic principle of brain function. *Nature Reviews Neuroscience*, *17*(12), 757.

Rogers, S. J., & Pennington, B. F. (1991). A theoretical approach to the deficits in infantile autism. *Development and psychopathology*, *3*(2), 137-162.

Rosenblum, L. A. & Andrews, M. W. (1994) *Influences of environmental demand on maternal behavior and infant development*. Acta Paediatrica Supplementum, 397, 57-63.

Roser, M. E., Fiser, J., Aslin, R. N., & Gazzaniga, M. S. (2011). Right hemisphere dominance in visual statistical learning. *Journal of cognitive neuroscience*, *23*(5), 1088-1099.

Rossouw, T. I. & Fonagy, P. (2012). Mentalization-Based Treatment for Self-Harm in Adolescents: A Randomized Controlled Trial, *Journal of the American Academy of Child and Adolescent Psychiatry, 51*(12). 1304-1313 https://doi.org/10.1016/j.jaac.2012.09.018.

Rossouw, T. I. (2015). The Use of Mentalization-Based Treatment for Adolescents (MBT-A) With a Young Woman With Mixed Personality Disorder and Tendencies to Self-Harm. *Journal of Clinical Psychology: In Session*, *71*(2), 178–187. https://doi.org/10.1002/jclp.22153.

Rutter, J. (2002). *Changing health behaviour: intervention and research with social cognition models*. McGraw-Hill Education (UK).

Saarni, C. (1984). An Observational Study of Children's Attempts to monitor their expressive Behavior. *Child Development, 55*, 1504-151. http://dx.doi.org/10.2307/1130020.

Sadler, L. S., Slade, A., & Mayes, L. C. (2006). Minding the baby: A mentalization-based parenting program. In J. G. Allen & P. Fonagy, *Handbook of Mentalization-Based Treatment*, 271-288. John Wiley & Sons, Ltd.

Salzman, C. D., & Fusi, S. (2010). Emotion, cognition, and mental state representation in amygdala and prefrontal cortex. *Annual review of neuroscience*, *33*, 173-202.

Sawyer, S. M., Afifi, R. A., Bearinger, L. H., Blakemore, S., Dick, B., Ezeh, A. C., & Patton, G. C. (2012). Adolescence: a foundation for future health. *Lancet*, *379*, 1630–1640. https://doi.org/10.1016/S0140-6736(12)60072-5.

Saxe, R., & Kanwisher, N. (2003). People thinking about thinking people: the role of the temporo-parietal junction in "theory of mind". *Neuroimage*, *19*(4), 1835-1842.

Saxe, R., & Powell, L. J. (2006). It's the thought that counts: specific brain regions for one component of theory of mind. *Psychological science*, *17*(8), 692-699.

Schafer, R. (1973). Concepts of self and identity and the experience of separation individuation in adolescence. *The Psychoanalytic quarterly, 42*(1), 42-59.

Schatz, C. J. (1992) The development brain. Scientific American, 267, 60-67.

Schatz, D. G., Oettinger, M. A., & Schlissel, M. S. (1992). V (D) J recombination: molecular biology and regulation. *Annual review of immunology*, *10*(1), 359-383.

Schore, A. N. (2018). The right brain implicit self: A central mechanism of the psychotherapy change process. In *Unrepressed unconscious, implicit memory, and clinical work* (pp. 73-98). Routledge.

Schore, A. N., & Newton, R. P. (2013). Using modern attachment theory to guide clinical assessments of early attachment relationships. In *Attachment-based clinical work with children and adolescents* (pp. 61-96). Springer, New York, NY.

Schore, J. R., & Schore, A. N. (2008). Modern attachment theory: The central role of affect regulation in development and treatment. *Clinical Social Work Journal*, *36*(1), 9-20.

Schore, J. R., & Schore, A. N. (2014). Regulation theory and affect regulation psychotherapy: A clinical primer. *Smith College Studies in Social Work*, *84*(2-3), 178-195.

Sebastian, C., Viding, E., Williams, K. & Blakemore, S. (2010). Social brain development and the affective consequences of ostracism in adolescence. *Brain & Cognition, 72*(1), 134-145. https://doi.org/10.1016/j.bandc.2009.06.008.

Seckl, J. R., & Meaney, M. J. (2004). Glucocorticoid programming. *Annals of the New York Academy of Sciences*, *1032*(1), 63-84.

Seminowicz, D. A., Shpaner, M., Keaser, M. L., Krauthamer, G. M., Mantegna, J., Dumas, J. A., & Naylor, M. R. (2013). Cognitive-behavioral therapy increases prefrontal cortex gray matter in patients with chronic pain. *The Journal of Pain*, *14*(12), 1573-1584.

Sharp, C., & Fonagy, P. (2008). The Parent's Capacity to Treat the Child as a Psychological Agent: Constructs, Measures Psychopathology. *Social Development*, *17*(3), 737–754. https://doi.org/10.1111/j.1467-9507.2007.00457.x.

Sharp, C., Fonagy, P. & Goodyer, I. M. (2006). Imagining your child's mind: Psychosocial adjustment and mothers' ability to predict their children's attributional response styles. *British Journal of Developmental Psychology, 24*(1), 197-214. https://doi.org/10.1348/026151005X82569.

Shear, M. K., Fyer, A. J., Ball, G., et al (1991) Vulnerability to sodium lactate in panic disorder patients given cognitive-behavioral therapy. *American Journal of Psychiatry*, 148, 795-797.

Sheppes, G., Suri, G. & Gross, J. (2015). Emotion Regulation and Psychopathology. *Annual Review of Clinical Psychology, 11,* 379-405. https://doi.org/10.1146/annurev-clinpsy-032814-112739.

Shing, Y. L., Lindenberger, U., Diamond, A., Li, S. C., & Davidson, M. C. (2010). Memory maintenance and inhibitory control differentiate from early childhood to adolescence. *Developmental Neuropsychology*, *35*(6), 679-697.

Shonkoff, J. P., Boyce, W. T., & McEwen, B. S. (2009). Neuroscience, molecular biology, and the childhood roots of health disparities: building a new framework for health promotion and disease prevention. *Jama*, *301*(21), 2252-2259.

Shore, L. M., & Tetrick, L. E. (1994). The psychological contract as an explanatory framework.' *Trends in organizational behavior*, *1*, 91-109.

Silbersweig, D., Clarkin, J., Goldstein, M., Kernberg, O., Tuescher, O., Levy, K., Brendel, G. et al. (2007). Failure of Frontolimbic Inhibitory Function in the Context of Negative Emotion in Borderline Personality Disorder. *The American Journal of Psychiatry, 164*(12), 1832-1841. doi: 10.1176/appi.ajp.2007.06010126.

Skarderud, F. (2007). Eating One' s Words: Part III Psychotherapy for Anorexia Nervosa — An Outline for a Treatment and Training Manual.

European Eating Disorders Review, *15*, 323–339. https://do i.org/10. 1002/erv.

Slade, A. (2004). The move from categories to process: Attachment phenomena and clinicalevaluation. *Infant Mental Health Journal, 25,* 269-283. https://doi.org/10.1002/imhj.20005.

Slade, A. (2005). Parental reflective functioning: an introduction. *Attachment & Human Development, 7*(3), 269-81. doi: 10.1080/14616 730500245906.

Slade, A. (2007). Reflective Parenting Programs: Theory and Reflective Parenting Programs: *Psychoanalytic Inquiry, A Topical Journal for Mental Health Professionals*, *26*(4), 640–657. https://doi.org/10.10 80/07351690701310698.

Smetana, J. G., Campione-Barr, N., & Metzger, A. (2006). Adolescent development in interpersonal and societal contexts. *Annual Review of Psychology, 57*, 255-284. Doi: 10.1146/annurev.psych.57.102904.190 124.

Sobesky, W. (1983). The Effects of Situational Factors on Moral Judgments. *Child Development, 54*(3), 575-584. doi: 10.2307/1130044

Solms, M., & Turnbull, O. H. (2011). What is neuropsychoanalysis? *Neuropsychoanalysis*, *13*(2), 133-145.

Somerville, L., Jones, R. & Casey, B.J. (2010). A time of change: Behavioral and neural correlates of adolescent sensitivity to appetitive and aversive environmental cues. *Brain and Cognition, 71*(1), 124-133. https://doi. org/10.1016/j.bandc.2009.07.003.

Sperber, D., & Wilson, D. (2004). Relevance theory. *Handbook of Pragmatics. Oxford: Blackwell*, 607-632.

Spiegel, D., Bloom, J., Kraemer, H. D., et al., (1989) *Effect of psychosocial treatment on survival of patients with metastatic breast cancer*. Lancet, ii, 888-891.

St Stoyanov, D. (2018). Psychiatry and neurolaw: An Essay on the Mind-Brain Problem and Legal Proof. *Balkan Journal of Philosophy*, *10*(1), 27-36.

Steele, H., & Steele, M. (1998). Attachment and Psychoanalysis: Time for a Reunion. *Social Development, 7,* 92-119. https://doi.org/10.11 11/1467-9507.00053.

Steele, H., & Steele, M. (2005). *The construct of coherence as an indicator of attachment security in middle childhood: The Friends and Family Interview*. In K. Kerns & R.

Steinberg, L. (2001). We Know Some Things: Parent-Adolescent Relationships in Retrospect and Prospect. *Trends in cognitive sciences, 9*(2), 69-74. https://doi.org/10.1111/1532-7795.00001.

Steinberg, L. (2004). Risk Taking in Adolescence: What Changes, and Why? *Annals of the New York Academy of Sciences, 1021*(1), 51-58. https://doi.org/10.1196/annals.1308.005.

Steinberg, L. (2005). Cognitive and affective development in adolescence. *Journal of Research on Adolescence, 11*(1), 1-19. doi:10.1016/j.tics.2004.12.005.

Steinwand, G. M. (1984). Adolescent individuation: the culmination of a developmental line. *The Journal of the American Academy of Psychoanalysis*, *12*(1), 43-57. https://doi.org/10.1521/jaap.1.1984.12.1.43.

Stern, D. N. (1985). *The Interpersonal World of the Infant: A View from Psychoanalysis and Developmental Psychology*. Karnac Books.

Stevenson, I. H., Rebesco, J. M., Miller, L. E., & Körding, K. P. (2008). Inferring functional connections between neurons. *Current opinion in neurobiology*, *18*(6), 582-588.).

Stortelder, F., & Ploegmakers-Burg, M. (2010). Adolescence and the reorganization of infant development: a neuro-psychoanalytic model. *The Journal of the American Academy of Psychoanalysis and Dynamic Psychiatry, 38*(3), 503-31. doi: 10.1521/jaap.2010.38.3.503.

Strathearn, L., & McClure, S. M. (2002). A functional MRI study of maternal responses of infant facial cues. In *Annual Scientific Meeting of the Society for Neuroscience*, Abstract Viewer (No. 517.5).

Strathearn, L., Fonagy, P., Amico, J. & Montague, R. (2009). Adult Attachment Predicts Maternal Brain and Oxytocin Response to Infant Cues. *Neuropsychopharmacology, 34*, 2655-2666. doi: 10.1038/npp.2009.103.

Sturesson, A. (2018). *Exercise-induced adult hippocampal neurogenesis and the effect of exercise and adult hippocampal neurogenesis on spatial learning and memory.*

Suomi, S. J. (1991) *Early stress and adult emotional reactivity in rhesus monkeys*. In Childhood Environmental and Adult Disease: Symposium No. 156 (ed. CIBA Foundation Symposium Staff), pp. 171-188. Chichester: Wiley.

Target, M. & Fonagy, P. (1996). Playing with reality II: the development of psychic reality from a theoretical perspective. *The International Journal of Psychoanalysis, 77*(3), 459-479.

Taubner, S., Kessler, H., Buchheim, A., Kächele, H., & Staun, L. (2011). The Role of Mentalization in the Psychoanalytic Treatment of Chronic Depression. *Psychiatry*, *74*(1), 49–58.

Thase, M. E., Fasiczka, A. L., Berman, S. R., et al., (1998) *Electroencephalographic sleep profiles before and after cognitive behavior therapy of depression*. Archives of General Psychiatry, 55, 138-144.

Thase, M. E., Greenhouse, J. B., Frank, E., Reynolds, C. F., Pilkonis, P. A., Hurley, K., & Kupfer, D. J. (1997). Treatment of major depression with psychotherapy or psychotherapy-pharmacotherapy combinations. *Archives of general psychiatry*, *54*(11), 1009-1015.

Thiel, A., Habedank, B., Herholz, K., Kessler, J., Winhuisen, L., Haupt, W. F., & Heiss, W. D. (2006). From the left to the right: How the brain compensates progressive loss of language function. *Brain and language*, *98*(1), 57-65.

Thiel, A., Schumacher, B., Wienhard, K., Gairing, S., Kracht, L. W., Wagner, R., & Heiss, W. D. (2006). Direct demonstration of transcallosal disinhibition in language networks. *Journal of Cerebral Blood Flow & Metabolism*, *26*(9), 1122-1127.

Trucco, E., Colder, C. & Wieczorek, W. (2011). Vulnerability to peer influence: A moderated mediation study of early adolescent alcohol use initiation. *Addictive Behaviors, 36*(7), 729-736. https://doi.org/10.1016/j.addbeh.2011.02.008.

Tyson, P. & Tyson, R. (1990). *Psychoanalytic Theories of Development: An Integration*. Ed. Yale University Press, New Haven & London

Uddin, L. Q. (2011). Brain connectivity and the self: the case of cerebral disconnection. *Consciousness and cognition*, *20*(1), 94-98.

Van IJzendoorn, M. H. (1995). Adult attachment representations, parental responsiveness, and infant attachment: a meta-analysis on the predictive validity of the Adult Attachment Interview. *Psychological bulletin, 117*(3), 387-403.

Van Overwalle, F. (2009). Social Cognition and the Brain: A Meta-Analysis. *Human Brain Mapping*, *30*, 829–858. https://doi.org/10.1002/hbm.20547.

van Praag, H., Zhao, X., & Gage, F. H. (2004). 9 *Neurogenesis in the Adult Mammalian Brain.*

Vanheule, S., Verhaeghe, P. & Desmet, M. (2011). In search of a framework for the treatment of alexithymia. *Psychology and Psychotherapy, 84*(1), 84-97. https://doi.org/10.1348/147608310X520139.

Viinamaki, H., Kuikka, J., Tiihonen, J., et al (1998) *Change in monoamine transporter density related to clinical recovery: a case-control study.* Nordic Journal of Psychiatry, 52, 39-44.

Wampold, B. E. (2012). Humanism as a common factor in psychotherapy. *Psychotherapy*, *49*(4), 445.

Wampold, B. E., & Carlson, J. (2011). *Qualities and actions of effective therapists*. Washington: American Psychological Association.

Winnicott, D. W. (1956). Primary maternal preoccupation. In *Collected papers, through paediatrics to psychoanalysis* (pp. 300-305). London: Tavistock Publications, 1958.

Winnicott, D. W. (1962). Ego-integration in child development. *In The maturational processes and the facilitating environment.* (p. 56-63). London: Hogarth Press and the Institute of Psycho-Analysis, 1965

Winston, J. S., Strange, B. A., O'Doherty, J. & Dolan, R. J. (2002). Automatic and intentional brain responses during evaluation of trustworthiness of faces. *Nature Neuroscience, 5,* 277-283. doi: 10.1038/nn816.

Wollheim, R. (1995). The mind and its depths. *Philosophical Quarterly, 45*(178), 100-103. doi: 10.2307/2219854.

Yeh, S.-R., Fricke, R. A. & Edwards, D. H. (1996) *The effect of social experience on serotonergic modulation of the escape circuit of crayfish.* Science, 271, 366-369.

Zachar, P., Stoyanov, D., Aragona, M., & Jablensky, A. (Eds.). (2015). *Alternative perspectives on psychiatric validation: DSM, ICD, RDoC, and Beyond.* International Perspectives in.

Zanardi, R., Barbini, B., Rossini, D., Bernasconi, A., Fregni, F., Padberg, F., & Bersani, G. (2008). New perspectives on techniques for the clinical psychiatrist: Brain stimulation, chronobiology and psychiatric brain imaging. *Psychiatry and clinical neurosciences*, *62*(6), 627-637.

Zikopoulos, B., & Barbas, H. (2006). Prefrontal projections to the thalamic reticular nucleus form a unique circuit for attentional mechanisms. *Journal of Neuroscience*, *26*(28), 7348-7361.

About the Authors

Vanya Matanova, PhD, DSc

Full Professor at Sofia University St. Kliment Ohridsky

Professor Vanya Matanova, full professor of clinical psychology at Sofia University. Head of the master's program "Clinical Psychology". Chairwoman of the Bulgarian Association of Clinical and Counseling Psychology and the Bulgarian Association of Dyslexia; president of the "Clinical Psychology in Action" foundation. Editor in chief of the "Clinical and Counseling Psychology" journal. Director of the Institute of Mental Health and Development, Sofia. Cognitive-behavioral psychotherapist. Specialized in Russia, UK, Ireland, Scotland, and Belgium. Guest lecturer in the UK, Macedonia, Cyprus, Kazakhstan. Thesis supervisor of 26 dissertations in the field of clinical psychology and neuro-psychology. Author of 12 books and over 200 articles, participant in national and international science forums.

Martin Kolev, PhD

Psychotherapist at the Institute for Mental Health
and Development, Sofia

Martin Kolev has a doctorate degree in clinical psychology, specializing in clinical work with children, adolescents and their families at the Institute of Mental Health and Development, Sofia.

Has a private practice as clinical supervisor in Mentalization-Based Treatment and has steadily been involved in training seminars for the Bulgarian Association of Cognitive-Behavioral Psychotherapy for the past decade.

Specialized in programs around the world including mainly England and Italy but also Greece, Switzerland, France, Wales. Authoring and co-authoring article publications and chapters of shared authoring studia in several European and American clinical journals. Main fields of interest include neuropsychology, mentalization-based practices of psychotherapy, childhood and adolescence development, transitions and crises, family functioning, psychodynamics of clinical work-ethics, therapeutic outcome measurements, reflective functioning and attachment.

Pia Tohme, PhD

Visiting Assistant Professor at Lebanese American University,
Beirut, Lebanon

Dr. Pia Tohme Khalaf graduated with a PhD from University College London, investigating parental mentalizing capacities with regards to their adolescent identical twins and its effects on attachment security. She is also trained in the Approach to Parenting Teenagers from the Open Door Young People's consultation services and in Mentalization Based Therapy for Children and Adolescents, as well as its application in a school setting, from the Anna Freud National Centre for Children and Families. Dr. Tohme Khalaf is a Visiting Assistant Professor of Clinical Psychology at the Lebanese American University and works in private practice with children, adolescents and their parents, based on attachment and mentalization principles. She is also a preschool consultant psychotherapist and conducts research at Father Andeweg Institute for the Deaf, focusing on theory of mind in young children and mentalizing in adolescents with hearing difficulties.

Zlatomira Kostova, PhD

Senior assistant at Plovdiv University Paisiy Hilendarsky

Zlatomira Kostova, works as an Assistant Professor at the Department of Psychology at the University of Plovdiv. MA in Clinical and Counseling Psychology since 2001. Interests in clinical psychology and individual and group cognitive-behavioral therapy. Licensed cognitive-behavioral psychotherapist. Activities include teaching disciplines in the field of clinical psychology, practical experience in clinical assessment, counseling and therapy with children and families.

Author of the monograph "Trauma and parenting'' (2018). Scientific interests in the field of clinical and counseling psychology. Active contribution "Mother's attachment as an adult Interview" (AAI); Practical and applied contribution related to development of individual profile for emotional-affective functioning and development in generalized developmental disorders.

Member of the Editorial Board of the Journal of Clinical and Counseling Psychology. Member of the Society of Psychologists in Bulgaria, the Association of Clinical Psychologists, the Bulgarian Association for Cognitive-Behavioral Psychotherapy, and the Bulgarian Ballint Society.

INDEX

α

β

A

B

C

D

E

F

G

H

I

R

S

T

U

V